T0338515

INTEGRATION, COORDINATION AND CONTROL OF MULTI-SENSOR ROBOT SYSTEMS

THE KLUWER INTERNATIONAL SERIES IN ENGINEERING AND COMPUTER SCIENCE

ROBOTICS: VISION, MANIPULATION AND SENSORS

Consulting Editor
TAKEO KANADE

Other books in the series:

Robotic Grasping and Fine Manipulation, M. Cutkosky
ISBN 0–89838–200–9

Shadows and Silhouettes in Computer Vision, S. Shafer
ISBN 0–89838–167–3

Perceptual Organization and Visual Recognition, D. Lowe
ISBN 0–89838–172–X

Three-Dimensional Machine Vision, Takeo Kanade
ISBN 0–89838–188–6

Object Recognition Using Vision and Touch, P. Allen
ISBN 0–89838–245–9

Robot Dynamics Algorithms, R. Featherstone
ISBN 0–89838–230–0

INTEGRATION, COORDINATION AND CONTROL OF MULTI-SENSOR ROBOT SYSTEMS

by

Hugh F. Durrant-Whyte
University of Oxford

KLUWER ACADEMIC PUBLISHERS
Boston/Dordrecht/Lancaster

Distributors for North America:
Kluwer Academic Publishers
101 Philip Drive
Assinippi Park
Norwell, Massachusetts 02061 USA

Distributors for the UK and Ireland:
Kluwer Academic Publishers
MTP Press Limited
Falcon House, Queen Square
Lancaster LA1 1RN, UNITED KINGDOM

Distributors for all other countries:
Kluwer Academic Publishers Group
Distribution Centre
Post Office Box 322
3300 AH Dordrecht, THE NETHERLANDS

Library of Congress Cataloging-in-Publication Data

Durrant-Whyte, Hugh F., 1961–
  Integration, coordination, and control of multi-sensor robot
systems / Hugh F. Durrant-Whyte.
    p. cm. — (Kluwer international series in engineering and
computer science ; SECS 36)
  Bibliography: p.
  Includes index.
  ISBN 0–89838–247–5
  1. Robotics. I. Title. II. Series.
TJ211.D87 1987   87–21479
629.8'92—dc19   CIP

Printed in the United States of America

To Daphne

Contents

List of Figures

Preface

Overview

Recent years have seen an increasing interest in the development of multi-sensory robot systems. The reason for this interest stems from a realization that there are fundamental limitations on the reconstruction of environment descriptions using only a single source of sensor information. If robot systems are ever to achieve a degree of intelligence and autonomy, they must be capable of using many different sources of sensory information in an active and dynamic manner.

The observations made by the different sensors of a multi-sensor system are always uncertain, usually partial, occasionally spurious or incorrect and often geographically or geometrically imcomparable with other sensor views. The sensors of these systems are characterized by the diversity of information that they can provide and by the complexity of their operation. It is the goal of a multi-sensor system to combine information from all these different sources into a robust and consistent description of the environment.

The intention of this book is to develop the foundations of a theory of multi-sensor integration. A central tenant of this theory is the belief that only through a quantitative description of sensor and system performance can we ever hope to understand and utilize the complexities and power provided by a multi-sensor system. A quantitative description of performance provides an a priori means of developing sensor capabilities and reasoning about the cumulative effects of observation uncertainty. We will maintain that the key to the efficient integration of sensory information is to provide a purposeful description of the environment and to develop effective models of a sensors ability to extract these descriptions.

We will develop a description of the robot environment in terms of a topological network of uncertain geometric features. Techniques for manipulating, transforming and comparing these representations are described, providing a mechanism for combining disparate observations. A general model of sensor characteristics is developed that describes the dependence of sensor observations on the state of the environment, the state of the sensor itself, and other sensor observations or decisions. A constrained Bayesian decision procedure is developed to cluster and integrate sparse, partial, uncertain observations from diverse sensor systems. Using this topological network describing the world model, a method is developed for updating uncertain geometric descriptions of the environment in a manner that maintains a consistent interpretation for observations. A team theoretic representation of dynamic sensor operation is used to consider the competitive, complementary, and cooperative elements of multi-sensor coordination and control. These descriptions are used to develop algorithms for the dynamic exchange of information between sensor systems and the construction of active sensor strategies.

Acknowledgments

Many people have made contributions to the ideas presented in this book. I would first like to thank Lou Paul, for his constant encouragement and support, and for his experience and patience in the face of overwhelming belligerence. I am greatly indebted to Max Mintz for his help and guidance, for his vast experience, and for acting as an almost limitless source of information. I would also like to thank "Our Fearless Leader", Ruzena Bajcsy, for providing the environment, motivation and spirit, without which this work would not have been possible. I also owe a great debt to Mike Brady for having the confidence to let me come and work in Oxford, for providing the facilities and the motivation to "do research".

I have benefited greatly from the contributions made by all the members of the Grasp Laboratory to this work and my stay at the University of Pennsylvania. In particular, I would like to thank Amy Zwarico for her moral support, Sharon Stansfield and Greg Hager for "discussions", Insup Lee (and his chain gang) for giving the world DPS, Alberto Izaguirre, Eric Krotkov, Dave Smitley, Franc Solina, Stephan Mallat, Gaylord Holder, John Summers, Jeff Trinkle, and Ray McKendall, for both their technical help and friendship.

I am grateful to the Trustees of the Thouron Award for giving me the opportunity to study at the University of Pennsylvania, and to the National Science Foundation for completing what the Thouron committee started. The research reported in this book was supported, in part, under NSF grants DMC-84-11879 and DMC-85-12838. Of course, all the conclusions and recommendations are my opinion, and are extremely unlikely to reflect the views (if any) of the National Science Foundation.

INTEGRATION, COORDINATION AND CONTROL OF MULTI-SENSOR ROBOT SYSTEMS

Chapter 1

INTRODUCTION

1.1   Sensors and Intelligent Robotics

Robots must operate in a world which is inherently uncertain. This uncertainty arises in the perception and modeling of environments, in the motion of manipulators and objects, and in the planning and execution of tasks. In present-day industrial robotics, this uncertainty is represented implicitly in programming and planning, requiring errors in operation to be small and well understand. Robust task execution is accomplished by increasing sensor and manipulator accuracy, using high tolerance and well known objects, and supplying parts in predetermined positions and orientations. This works well in situations where the environment, tasks and motions are all completely determined in advance. However, as robot systems are increasingly required to operate in more complex domains, where the environment and tasks are uncertain or unknown, then this implicit allowance for uncertainty will be insufficient to ensure efficient operation.

If robots are to extend their abilities to operate in uncertain environments, sensor systems must be developed which can dynamically interpret observations of the environment in terms of a task to be performed, accounting for this uncertainty and re-

1

fining the robots model of the world. Current robotics research has concentrated on using single sensors to extract particular features of an uncertain but known environment. In general, these systems make observations of simple features (line segments for example), and by matching these to a data base of objects, attempting to determine the location and orientation of objects [Bolle 86, Grimson 84, Lowe 85]. These single sensor systems work well in situations where the environment is structured and objects are well known. However these single-sensor single-algorithm systems are severely limited in their ability to resolve ambiguities, identify spurious information and to detect errors or failure. These shortcomings are *not* a product of the algorithms employed, they are an *unavoidable* consequence of attempting to make global decisions based on incomplete and underconstrained information.

Single sensor systems are only capable of supplying partial information and are generally limited in their ability to resolve ambiguities in unknown or partially unknown environments. If robot systems are to operate in unstructured or uncertain environments, they must make use of multiple sources of sensory information. By combining information from many different sources, it would be possible to reduce the uncertainty and ambiguity inherent in making decisions based on only a single information source. The use of many physically distinct information sources additionally allows the development of distributed, highly parallel observation processing, achieving more rapid data extraction and providing partially redundant, robust systems operation [Giralt 84].

The sensors of a multi-sensor robot system are characteristically complex and diverse. They supply observations which are often difficult to compare or aggregate directly, and which can be both geographically and geometrically disparate. These observations are usually partial, always uncertain and often spurious. It is the goal of a multi-sensor system to combine information from all these different sources into a robust and

consistent environment description. To make efficient use of
sensory information, we must model the capabilities of each
sensor to extract information from the environment, and to
provide a means of communicating this information in a common framework. This description should provide a means of
distributing decision making and integrating diverse opinions.

We maintain that the key to efficient integration of sensory
information is to provide a purposeful description of the environment and to develop effective models of a sensors ability to
extract this information. The environment description should
provide a homogeneous description of different observable features, allowing for the manipulation and transformation of different descriptions between different sensory systems. These
descriptions should take explicit account of the uncertainty involved in sensor interpretation, and provide a means to communicate partial hypotheses to other sensors. The model of sensor
capabilities must also take explicit account of uncertainty, but
must further describe the dynamic nature of sensory information; how information is communicated and how it can be used
to guide observations and resolve or interpret different sensor
hypotheses.

In this book, we will consider the problem of integrating information from different sensory sources. We will concentrate
on four basic issues; how to organize a distributed sensing system, how to integrate diverse sensor observations, how to coordinate and guide the decisions made by different sensors, and
how to control devices toward improving sensor system performance. We will maintain a philosophy of requiring a quantative description of the sensing process, providing an a priori
estimation of system performance. This will allow the development of robust integration strategies and efficient policies of
coordination and control. We will emphasize the importance
of modeling, both of the environment and the capabilities of
sensing systems. Modeling is essential to understanding the
complexity and utilizing the power provided by multi-sensor

systems.

1.2   Multi-Sensor Robot Systems

Recent years have seen an increasing interest in the development of multi-sensor systems [Henderson 87]. This interest stems from a realization of the fundamental limitations on any attempt at building descriptions of the environment based on a single source of information. Single sensory systems are only ever capable of supplying partial information and are consequently limited in their ability to resolve interpret unknown or partially unknown environments: If robot systems are ever to achieve a degree of intelligence and autonomy, they must be capable of using many different sensors in an active and dynamic manner; to resolve single sensor ambiguity, to discover and interpret their environment.

A multi-sensor robot system comprises many diverse sources of information. The sensors of these systems are characterized by the variety of information that they can provide and by the complexity of their operation. The measurements supplied by the sensors are uncertain, partial, occasionally spurious or incorrect and often geographically or geometrically incomparable with other sensor views. This sensor information is inherently dynamic; the observations supplied by one sensory cue may depend on measurements made by other sensors, the quality of the information provided may depend on the current state of the robot environment or the location and state of the sensor itself. It is the goal of the robot system to coordinate these sensors, direct them to view areas of interest and to integrate the resulting observations into a consistent consensus view of the environment which can be used to plan and guide the execution of tasks.

In order to make efficient use of sensory information, it is important to model the environment and sensors in a manner which explicitly accounts for the inherent uncertainty encountered in robot operation [Durrant-Whyte 86b]. We will consider a world which is described in terms of geometry. A model of a task and associated working environment can be developed in terms of the location and geometry of objects [Paul 81]. We maintain that the uncertainty intrinsic to operation in the real world should be represented explicitly in the geometry describing the environment. We will develop a model of the environment based on stochastic geometry [Kendall 61], representing geometric features as constrained functions. The uncertainty in each feature is described by a probability density function defined on the parameter vector. We show how these uncertain geometric objects can be manipulated and transformed in an efficient and consistent manner. This uncertain geometry, can provide a powerful ability to describe and reason about the effects of uncertainty in sensing and manipulation.

We consider sensors in terms of their ability to extract uncertain descriptions of the environment geometry. There are a variety of sensors that can be used to extract geometric observations of the environment; vision systems, tactile sensors, ultra-sound rangers, for example. A single sensing device can often apply a number of different algorithms or cues to its raw measurement data in order to extract particular types of geometric feature. For example, a stereo camera can recover depth from vergence or focus [Krotkov 87], or edges and surfaces by segmentation [Ballard 82, Horn 86]. The manipulation and transformation of uncertain geometric observations provides a basis for the communication of information between different sensor systems and allows the development of techniques to compare and combine disparate observations.

We will develop a team-theoretic model of sensor abilities. The purpose of a sensor model is to represent the ability of a sensor to extract descriptions of the environment in terms of a prior world model. Sensor models should provide a *quantitative* ability to analyze sensor performance, and allow the development of robust decision procedures for integrating sensor information. We can divide a sensor model into three parts; an *observation* model, a *dependency* model, and a *state* model. An observation model is essentially a static description of sensor performance, it describes the dependence of observations on the state of the environment. The construction of observation models is a well developed but still open research issue. In the case of a robot sensor system, observation models are made more difficult because of the often complex way in which the primitive data is built up in to measurements of geometric features. A dependency model describes the relation between the observations or actions of different sensors. For example, the observations made by a tactile probe may depend on the prior observations supplied by a vision system, or an edge detector may provide a segmentation algorithm with a first estimate of region boundaries. A state model describes the dependence of a sensors observations on the location or physical state of a sensing device. For example, a mobile camera platform, may be able to change its location (viewpoint) or the focus of its lens to provide different observation characteristics. A description of the dependency of observations on sensor state would enable the development of sensor strategies.

Sensor information is inherently dynamic; observations provided by one sensory cue cannot be considered in isolation from the observations and actions of other sensors. Different sensors may provide quite disparate capabilities, which when considered together are *complementary*. If we have a number of sensors that provide essentially the same information for the purpose of reducing uncertainty and allowing redundant operation, we are likely to have information that is *competitive*, or in disagreement. When two or more sensory cues depend on each other for guidance (multiple-resolution edge detection, or a pre-attentive visual process, for example), then these information sources must *cooperate* to provide observations.

If we are to provide a means to analyze of information in

a multi-sensor system, we must develop an effective model of a sensors capabilities, with which to provide an analytic basis for making quantative decisions about sensor organizations. We will develop a probabilistic model of sensor capability in terms of an *information structure*. We consider a multi-sensor system as a team, each observing the environment and making local decisions, each contributing to the consensus view of the world, and cooperating to achieve some common goal. The information structure of a team describes the observations and decisions made by the team members, how they are related, and how they depend on the organization and operation of the team. The concept of a team of sensors provides a powerful means to describe and analyze multi-sensor systems.

1.3 Organization of Sensor Systems

The increased complexity of tasks and environments in which robot systems must work requires that many more devices and computational resources be used to achieve a specified goal. This increased resource requirement forces complex robot systems to be distributed amongst many different processes [Ruoff 85]. This distribution may take the form of an increased number of physical devices (sensors, manipulators, etc) or increased computational resources (multi-computer, parallel machines, etc). This is particularly true in a multi-sensor robot system were devices are often geographically as well as logically distinct and the algorithms used to interpret sensor information are complex and require large computational resources to achieve a specified level of performance. A similar kind of problem is encountered in the more general setting encountered in distributed knowledge based systems [Ghallab 85, Ginsberg 85]. In this case, there is clearly a limit to the amount of processing or reasoning any single agent or device can achieve. Recently, there has been a great deal of research in distributed artificial intelligence (DAI) [Corkill 79, Davis 83, Lesser 81, Smith 81]. This work has focused on problems of organization [Chandrsekaran 81], communication [Kornfeld 81] and the distribution of reasoning tasks [McArthur 82]. There has however been very little work on the problem of integrating information from distributed agents; combining different opinions, resolving conflicts and finding a consensus decision. One problem with most DAI techniques is that the heuristic basis for their operation often precludes an analysis of the decision making process. This in turn makes them difficult to extend consistently or to use them in different situations.

The need to distribute problem solving and task execution ability gives rise to three basic problems; how best to organize a complex distributed system [Kleinrock 85], how to distribute or allocate tasks amongst the agents of the system [Smith 80], and how to integrate the various opinions from different agents [Rosenschein 82]. There is no clear way to answer any of these questions, and they remain the subject of a great deal of current research. In the case of a multi-sensor robot system, we have many physically distinct devices that must operate in an environment that is inherently uncertain. These systems must provide for fast, robust and extensible operation. They should achieve a high degree of parallelism, allowing as much local control as possible [Paul 85], distributing decision making and task planning amongst many computational choices.

The importance of structure in complex systems can not be overstated [65]. Even in relatively well defined problems there is justified disagreement on how complex systems should be organized. These problems have received much attention in the literature particularly from economists and corporate strategists [Galbraith 73]. In organization theory the limit of a single distinct system to consider complex functions in isolation is termed "bounded rationality" [McGuire 72, Simon 62]. There have been a number of proposals made as to what organizational form a robot system should take. These can be approximately divided into three categories: Hierarchical, heterarchical and layered systems.

Hierarchical control systems are rigidly structured organizations in which a supervisor decomposes high level goals in to more and more primitive subtasks until the point is reached where the subtasks are small enough to apply as individual actuations in mechanical members or sensor systems [Brooks 85]. Hierarchical robot control systems are by far the most popular organizational structures to be used [Dufay 84]. The STRIPS planner [Fikes 71] is typical of this approach. The main advantage of this type of organization is the ease with which large complex problems can be decomposed into simpler subtasks in a way that makes coordination easy to achieve. However, a pure hierarchy precludes direct interaction with low-level sensor and action processes at higher levels and tends to result in a very strict structured decomposition of problems, leading to overall longer processing times and poor use of sensory information. Further, this sequential structure causes the robot to rely on one system after another to define task execution, a single error in a single system totally disorients the robot.

In contrast to a hierarchy, a heterarchical controller [Minsky 72, Winston 77] allows each individual element of the control system to be considered equally. Communication is allowed between any two elements and there is no overall supervision of the task. One advantage of this type of organization is that knowledge can be integrated at any level in the most useful way, obtaining maximum use of sensor information. However, the control of such an amorphous structure is all but impossible. It is not difficult to see that chaotic actions may result. The nearest realization to such a structure has been the Hearsay speech understanding system [Erdman 79, Lesser 79], in which multiple independent knowledge sources communicate to each other via a common message passing and world modeling facility called a Blackboard. The Blackboard concept forms the basis for many less heterarchical organizations presented in the robotics literature [Elfes 84, Harmon 84]. The reason that the Blackboard structure is so popular is because of its flexibility to additional subsystems, and its robustness to the effects of uncertainty and error in distributed knowledge based systems. The heterarchical concept has also given rise to many more structured organizations based around the idea of distributed knowledge sources with loose communication facilities.

An interesting structure that lies somewhere between a hierarchy and a distributed knowledge source organization is the so-called layered control system. Arbib [Arbib 71] introduced the idea to the computer science literature, where he showed that the perceptual system of a frog has a layered structure called a somatotopy. This layered organization was shown to be very efficient for perceptual functions. More recently Brooks [Brooks 84,85,86] has suggested a control organization layered by behavioral function. This constitutes more of a vertical distribution of control in contrast to more traditional horizontal layering. The advantage of this is that incremental behavior additions can be made. However, this can only be achieved at the cost of making functional additions more difficult.

We can increase the overall robustness of the system by avoiding overdependence in a hierarchical structure, allowing loosely coupled communication facilities at any single layer. A useful concept that satisfies requirements of minimized but structured communication is that of having strategic and tactical levels of control [Orlando 84]. This is essentially a two level organization. The strategic level looks at long range planning, coordination and exception handling. The tactical level allows loose communication among its functional units and implements policy in the short and medium term. Organizational theory also suggests such a structure [Galbraith 73], where the functional elements of the system are essentially self-contained and self-organizing, only deferring to a hierarchical structure (strategic level) to handle coordination and exceptions. This type of organization minimizes communication by only utilizing the strict hierarchical structure when coordination is required, and otherwise maintains a very loosely coupled and flexible structure to which both additional functional and behavioral elements may be added.

We propose that a robot system be organized as a team of expert agents implementing tactical plans and being supervised by a strategic executive. We maintain that this organization imposes sufficient structure coordination and exception handling while admitting a loosely coupled communication to be maintained during normal execution. We will develop an organizational structure which accounts for this uncertainty and allows the sensor agents to make mistakes or supply spurious information. The complexity of many sensor algorithms suggests that we should also distribute as much local decision making as possible to the sensor systems. This in turn allows the agents of the system to be loosely coupled, providing increased robustness and extensibility. We maintain that because of the diverse nature of sensor observations, the sensor agents should be allowed to maintain their own local world model, specific to local sensing requirements, and that sensors should be endowed with as much local decision making ability as possible. Each of the distributed agents is expert in its own localized function and contains its own localized knowledge source and agent-specific world model. We maintain that minimal communication between team members or agents will be achieved when distribution is performed by function alone. That is each agent is associated with some function, vision, touch, manipulation, etc. In this way the agents can maintain very specific functional models and algorithms, only communicating information that is relevant to the rest of the system.

1.4   The Integration of Sensory Information

Robot sensors can provide very diverse observations of the environment. These observations are always noisy and uncertain, usually partial, sparse or incomplete, and occasionally spurious. To utilize multiple sensory information effectively we must provide a robust, efficient and consistent methodology for integrating disparate observations. The key to developing efficient integration methods is to describe the environment in a consistent and homogeneous fashion, and to provide an efficient representation of a sensors ability to extract this information. We shall only be concerned with the geometry of the environment, (edges, surfaces, centroids, etc), and provide for uncertainty by explicitly representing a feature by a probability distribution on its associated parameter vector. We maintain consistency in the interpretation of uncertain features by requiring the parameter vectors to maintain the natural topology of the environment. We will describe the observations from a particular sensor by an information structure. The information structure of a sensor describes the dependence of the sensors observations on the state of the environment (observation model), the other sensor decisions (dependency model) and the state of the sensor itself (action model).

This probabilistic description of the environment and sensor capabilities provides a framework in which to apply the tools of statistical decision making to multi-sensor robot systems. We will develop general techniques for representing uncertain geometric features and manipulating them in a consistent manner. We embed the information structure of each sensor in a team-theoretic framework. This in turn allows the development of information dependency models and procedures for guiding sensor observations. We will utilize Bayesian clustering techniques to filter observations and develop a framework for the consistent integration of sensor information in to a prior world model.

1.5   Coordination of Sensor Systems

The dynamic nature of sensor information is particularly important in a multi-sensor robot system. We want to be able to coordinate information from different sensors so that we can guide the observations of the environment in a purposeful manner. There are two main elements to the coordination of sensory information, the guiding of a sensors observations based on previous sensor measurements, and the process of relocating or changing the state of active sensor devices to improve the observations that they provide.

Robot sensor systems generally supply only very sparse and incomplete observations [Grimson 84]. A major advantage of multi-sensor systems is to alleviate the problems involved in recognition using sparse date by having many sensors suggesting and verifying single sensor hypothesis [Terzopoulos 86]. Generally the observations provided by a single sensor can be given many different and locally consistent interpretations. Using observations from another sensor, the possible interpretations can be reduced. Using many sensors, we can eventually resolve this to a single interpretation. In this way, it is possible to reduce the natural ill-posedness of making decisions based on single sensor observation sets, by using more information from other sensors, rather than applying artificial prior information to reduce the search space. The verification of single sensor hypothesis by comparison with other other sensor observations also provides a mechanism with which to reduce uncertainty and eliminate spurious observations. If we are to fully utilize a sensors ability, we must develop techniques to coordinate this transfer of information. This dependency between information sources leads to the idea of networks of sensors and cues, engaged in a dialogue, successively localizing features in the environment and resolving ambiguity between each other. To develop and realize the structures that result from this type of interaction we must provide some basis for the analysis of information structures.

We propose a team-theoretic formulation of a multi-sensor system in the following sense: The sensor agents are considered as members of the team, each observing the environment and making local decisions based on the information available to them. A manager (executive or coordinator) makes use of utility considerations to converge the opinions of the sensor systems. Teams can be viewed as organizations of interacting experts, each with complimentary, overlapping or cooperative opinions. A team-theoretic characterization of sensor agents and manager provides a common framework for the comparison of diverse observations and the coordination of different sensor actions.

1.6   Summary and Overview

The work presented in this book is intended to develop the foundations for a theory of multi-sensor systems. We will provide a methodology for the integration of partial, diverse, uncertain sensor information into a robust and consistent estimate of the state of a robot environment. We will develop techniques which allow the communication and utilization of information from different sources, accounting for the dynamic nature of sensor observations in resolving descriptions of the world.

We will maintain that the key to the successful integration of disparate sensory information is to provide a purposeful description of the environment and an efficient model of a sensors ability to extract these descriptions. Chapter 2 develops a general environment model. We restrict our attention to purely geometrical descriptions of the environment and describe each feature as a function together with a parameter vector on which is defined a probability density function. Techniques are developed to manipulate, transform and combine these uncertain features, providing a method for the comparison of sensory information. A key element in this formulation is to develop an underlying network of objects, features, sensors and cues in the robot environment. This network describes the topology associated with the environment and provides a mechanism for updating and comparing information in the world model while maintaining a consistent interpretation of the environment. The network structure underlies many of the integration algorithms that will be developed and is seen as the keystone in providing an efficient and consistent sensor integration policy.

Chapter 3 introduces a team-theoretic model of sensor capabilities. These models represent a sensors ability to extract uncertain geometric descriptions from the environment. We will develop a model of a sensors ability to observe the environment, in terms of the state of the environment, the dependence of it's observations on the measurements and actions of other sensory systems, and on the location or state of the sensing device itself. This model provides a powerful description of the competitive, cooperative and complimentary nature of dynamic sensor interactions.

Chapter 4 takes these environment and sensor models and applies them to the problem of observing and updating a model of a partially known environment. Bayesian techniques are developed to cluster observations and provide a robust estimator of environment states. The natural topology of the environment is described in terms of a stochastic network with arcs denoting relationships between geometric objects. To integrate observations into a world model in a consistent manner, the topology of the environment and the structure of this network must be maintained. We will develop techniques for updating a general network of this form to provide a consistent interpretation for observations and to propagate information amongst related geometric objects.

Chapter 5 makes use of the team theoretic description of sensing capabilities to extend these integration procedures to sensors taking disparate observations in a dynamic, interacting manner. We discuss and develop methods for dealing with overlapping or complimentary information sources, competitive information or disagreement between sensor systems and special cases of cooperative behavior between different sensory sources. Using the model of sensor dependence on state, maximal-information sensor control strategies are developed.

Chapter 6 describes the implementation of these ideas on a distributed sensing system, using a mobile stereo camera system, with three visual cues, and a tactile-force array, as sensor agents.

The primary purpose of this book is to provide a theoretical framework in which to develop methods for the coordination, control and integration of multiple sensor robot systems. The key elements of this theory are:

- The consistent representation of uncertain geometry and the development of techniques to manipulate, transform and propagate uncertain geometric information.

- The use of topologically invariant properties of the environment to provide a consistent interpretation of sensory information.

- The construction of efficient algorithms for integrating sparse, partial, uncertain observations from multiple disparate sensor systems.

- The development of sensor models describing the dependence of observations on environment state, sensor location and other sensor decisions.

- The use of these models in providing a mechanism for the dynamic exchange of sensory information and the development of active sensing strategies.

The theory and techniques developed in this book provide a foundation on which the study of multi-sensor systems can proceed.

Chapter 2
ENVIRONMENT MODELS AND SENSOR INTEGRATION

2.1 Introduction

The capabilities of a robot system are strongly dependent on an ability to observe and understand the environment in which it operates. For this reason, an effective and purposeful description of the environment is fundamental to the development of intelligent robotics [Ambler 86]. There are many different ways in which the problem of description can be approached, resulting in a variety of different environment modeling techniques; solid models, primitive surface models, or functional descriptions, for example. The utility of each description depends on the specific task in which it will be used. Ideally a robot system should provide many different levels of environment description, and be able to move freely and maintain consistency between them.

We will be concerned with the problem of integrating observations of the environment from many different sensory sources. The ability of a sensor to extract descriptions from the envi-

19

ronment is, in part, the dual of the world modeling problem. In the most general sense, a sensor provides a mechanism to extract or reconstruct descriptions of the environment, described by the observed information, in terms of a prior world model [Poggio 84]. The relation between sensor observations and the robot's description of the world is fundamentally important: The ability to consistently model the environment and manipulate object descriptions provides a mechanism with which to compare and integrate sensor observations. Indeed the representation of an environment maintained by the robot's world model provides a basis for the communication of information between different sensory cues. We will maintain that geometric primitives, of some form, are a natural environment description for sensor-based robotics: Each sensory cue can be considered as providing the robot system with observations or estimates of a specific type of geometric primitive; edges, surface normals, etc. This information can be communicated to other sensors, observing different types of features, through geometric transforms. These primitive geometric elements can then be compared and aggregated by the robot system, to provide a full and complete environment description, and subsequently used in the planning and execution of tasks.

The geometric elements used for the description and communication of information should consist of object features readily extracted by a single sensory cue; lines, surfaces, depth constraints, etc. Any higher-level description of the environment, in terms of form or function, would require more information than any single sensor could reasonably supply, and as a consequence would force systems to be rigidly organized. Conversely, any lower-level description, in terms of an irradiance image or depth map, provides a plethora of information detail, useful to only single sensor systems. As a consequence, these models inhibit any distributed decision making philosophy and increase the requirement for communication bandwidth.

Using geometric primitives at a sensor system level does

not, however, preclude the use of other models at *appropriate* levels within other parts of the robot system; functional models are important when the sensed environment is to be interpreted, irradiance models are required to process and extract geometry from the raw sensor data. Indeed, multiple resolutions of model are essential to a complete system, and must be integrated effectively with the geometric primitives used for describing sensor operation.

The geometric elements used in sensor processing must take explicit account of the uncertainty inherent to their operation. We must be able to manipulate these descriptions, intersect and aggregate them, and be able to move representations between coordinate systems. Understanding these uncertain geometric descriptions will not only supply an effective environment model but also form a basis for integrating disparate sensory information [Bolle 86].

We maintain that an effective and consistent description of geometry is fundamental to the efficient operation of sensor systems. It provides a mechanism with which to describe observations, it is the basis for integrating different measurements, and it is the language by which sensors communicate information to each other. We therefore feel justified in presenting a detailed development and description of uncertain geometry.

In the following we shall present a probabilistic viewpoint of uncertain geometry. We maintain that uncertainty should be represented as an intrinsic part of all geometric features. The representation of uncertain points, lines, curves and surfaces is developed by considering geometric features as stochastic point processes or random vectors described by a probability distribution on the parameter space of the associated object. The manipulation of uncertainty measures then becomes the transformation and combination of probability distributions. This representation provides a mechanism for the efficient manipulation of geometric uncertainty measures and a methodology for providing a consistent interpretation of geometric objects in terms of an invariant stochastic topology. This topology describes the relationships between geometric objects that do not change when the environment is uncertain. We develop the idea of an invariant measure over the group of transforms appropriate to Euclidean space (descriptions of geometric uncertainty that do not change when transformed in Euclidean space) and show that the Gaussian distribution is a reasonable approximation to this. Using a Gaussian-feature assumption it is shown how uncertain geometry can be manipulated in an efficient and consistent manner. We show that invariant relations between geometric objects can be considered as invariant elements of a stochastic topological network. This allows the consistent composition or propagation of geometric uncertainty, providing a mechanism with which to reason about the effects of uncertain geometric descriptions.

Section 2.2 reviews the development of geometric environment models in robotics. Section 2.3 suggests that a representation sympathetic to sensor systems should take explicit account of geometric uncertainty. Section 2.4 develops a representation for uncertain geometry in terms of constrained, parameterized functions. Each function describes a particular type of geometric feature, and each specific parameter value represents an instance of this feature type. Uncertainty is described by placing a probability distribution function on the parameter vector of a given feature type. The properties of these descriptions are discussed, and techniques for building consistent representations of geometric objects and relations are discussed. Section 2.5 describes how these uncertain representations can be manipulated and transformed between coordinate systems. Section 2.6 develops techniques to describe changes in feature location and description when the probability distribution on parameter vectors is assumed to be Gaussian. Section 2.7 demonstrates how constraints between descriptions can be used to maintain consistency amongst uncertain geometric features.

This development of uncertain geometry is fundamental in providing a firm basis for the description of sensor systems. Subsequent Chapters will use these techniques in the development of sensor integration strategies.

2.2 Geometric Environment Models

There is considerable debate as to what world model is most appropriate for a robot system engaged in sensing and general manipulation [Boissonnat 85]. We restrict our interest in world modeling to purely geometric descriptions of the environment. The problem of building a geometric model of the robot environment has received considerable attention (see for example [Ambler 75, Brady 82, Pentland 85, Requicha 80]). There are essentially two classes of geometric representations, the functional model and the image model.

Functional models derive from the need to solve problems in planning, assembly, computer aided design (CAD), and other high-level robot operations. There is a great variety of functional models in use, usually specialized for a particular task or to solve a specific representation problem (generalized cylinders, or solid geometry, for example). In general the CAD approach to environment and object modeling is to utilize some subset of constructive solid geometry (CSG) [Shafer 85]. It is rare that these models provide any mechanism for the representation of uncertainty in their descriptions. This consideration often precludes their use in sensory driven robot operations where uncertainty is intrinsic to problem solution., and where elemental descriptions are too global to allow single-sensor decision making.

Conversely, models of geometry as images; relations between small elemental image parts, and stochastic or Markovian random field models (MRF), provide a view of geometry as an array of simple scaler components, with no attempt to provide a functional description. These image models provide a great deal of statistical micro-structure from which uncertain descriptions of the environment can be reconstructed [Terzopoulos 84, Marroquin 85]. However they do not provide the functional representation or level of abstraction that we require to develop a multi-sensor paradigm, nor do they allow task level descriptions of the environment.

There is clearly a need for some compromise between these two extremes: We require a level of representation which can generaly allow the development of geometric models for specific tasks but which takes explicit account of the intrinsic uncertainty introduced by sensor driven recognition and task execution. Ambler [Ambler 86] has suggested that the representation of objects in a robot system's world model should be sympathetic to the nature of the sensor systems observations of the environment. Such a "representation for sensing" results in a surface and feature based description of objects. Within a system consisting of many diverse sensor agents, different models of the environment will be appropriate for different types of sensory information. We maintain that the overall representation of the world model should encourage a homogeneous description of all types of geometric features observed by sensors. This representation should provide for the description of uncertainty and partial information, and be capable of supporting, manipulating and comparing a variety of geometric sensor primitives including their intersections, relations and extents. Further, we will support a continual aggregation and compression of information, from local sensor image descriptions, to surface representations, to solid models. This aggregation increases the robustness of object descriptions at each stage, enables parallel fusion procedures to be used, increasing the flexibility of the system as a whole to the use of multiple disparate sensor systems.

2.3   Uncertain Geometry

We maintain that the key to both environment and sensor modeling is to find an efficient representation of uncertain geometry and to develop effective methods for manipulating, comparing and combining uncertain geometric structures. The need for a representation of geometric uncertainty has been recognized in the literature. Some notable examples occur in planning [Brooks 82, Chatila 85, Taylor 76], object recognition [Grimson 85] and navigation [Brooks 84, Smith 86].

The general policy in all these representations is to assume error bounds on geometrical parameters, resulting in "error manifolds" and "uncertainty cones", over which the error is implicitly assumed to be uniform. The problem with such measures of uncertainty is that they are difficult to manipulate and reason with in a consistent manner. It is generaly not sufficient to consider uncertainty in a single coordinate system, we must also be able to reason about the effects of error in coordinate frames other than those in which the uncertainty measure was initially defined. In this case, simple geometric objects describing uncertainty in one coordinate system may no longer be simple in another frame of reference. This problem is further compounded when any sequence of uncertainty measures must be aggregated to study their cumulative effect. Problems also arise when two or more different features must be compared, how for example, should an uncertainty cone be combined with an error manifold, and how does the result affect objects described in other coordinate systems? In a statistical sense there are complications resulting from the combination of two or more uncertainty measures; are the initial measures independent, is the assumption of uniform distributions appropriate, what is the combined distribution of the uncertainty measures and how should partial information (infinite uncertainty in one or more degrees of information) be represented?

To answer these questions we require a method which can

represent uncertain geometric features in a consistent, homogeneous manner, which allows these uncertainties to be manipulated easily, and which provides an ability to reason about the effects of uncertainty in the planning and execution of tasks.

2.4   Characterizing Uncertain Geometry

In section 2.4.1 we introduce a probabilistic description of uncertain geometry described in terms of implicit parameterized functions together with a probability distribution defined on the parameter vector. This provides a homogeneous framework for describing all the elements of classical geometry. We note the relation between this framework and the theory of random sets. Section 2.4.2 identifies certain desirable properties of these parameterizations, outlining a procedure for verifying that a specific representation of a geometric element is statistically well-behaved. This leads directly to the development of general mechanisms for transforming geometric elements between coordinate systems and for aggregating or intersecting different geometric objects. Section 2.4.3 develops the idea of stochastic invariance, in which geometric properties or relations remain unchanged when they are made uncertain. These invariants are closely related to classical topological invariants, so that we shall call the study of stochastic invariance, stochastic topology. The importance of stochastic topology lies in its ability to provide a consistent interpretation for uncertain geometric objects, and as a consequence, to act as a constraint system through which geometric information can be propagated.

2.4.1 Stochastic Geometry

All geometric objects (features, locations and relations), can be described by a parameter vector p and a vector function

$$g(x, p) = 0; \qquad x \subseteq \Re^n, \quad p \in \Re^m \qquad (2.1)$$

This function can be interpreted as a model of the physical geometric object, that maps a (compact) *region* $x \subseteq \Re^n$ in Euclidean n-space to a *point* $p \in \Re^m$ in the parameter space. Each function g describes a particular type of geometric object; all straight lines, or the family of quadratic surfaces, for example. Each value of p specifies a particular instance of a geometric object modeled by g. For example, *all* plane surfaces can be represented by the equation;

$$g(x, p) = p_x x + p_y y + p_z z + 1 = 0 \qquad (2.2)$$

with $x = [x, y, z]^T$ and $p = [p_x, p_y, p_z]^T$. A specific plane can be represented as a point $p \in \Re^3$ in the parameter space.

Consider a sensor observing a particular type of geometric object in the environment. The sensor can be considered as observing values of p. The uncertain *event* that the sensor "sees" a specific instance of this geometric object can be described by taking the parameter vector p to be a random variable. The likelyhood that a particular instance of a geometric feature is observed can now be described by a probability density function (p.d.f.); $f_g(p)$. For a specific type of feature represented by Equation 2.1, this p.d.f. describes the probability or likelyhood of observing a particular instance of the associated geometric object. It should be noted that this is *not* a noise model.

In this context, the analysis of uncertain geometry can, in theory, be reduced to a problem in the transformation and manipulation of random variables. The study of random geometry (stochastic geometry, geometric probability)[1] has a long

---

[1] Also integral geometry [Santalo 76].

and illustrious history. It differs from conventional probability in requiring a physical interpretation to be placed on random variables, resulting in physical (geometric) constraints on functions and relations. It is this difference which makes stochastic geometry a more difficult problem than just the manipulation of random variables. The classic example of this situation is the ubiquitous Bertrands Paradox [Bertrand 07], in which the probability of a random line intersecting a circle is shown to depend on the definition of 'line' and 'circle'. Such problems have resulted in random geometry being largely neglected as a sub-discipline of random variable theory. There have, however, been some important contributions to the subject in recent years.

In their book, Kendall and Moran [Kendall 63], describe a method of choosing distributions on geometric elements which provide a consistent interpretation of physical geometric elements. Although they concentrate on the distributions of points in Euclidean space (geometric probability), some important properties of stochastic lines and planes are also described.

In his thesis, Davidson [Davidson 68] made the important observation that arbitrary random geometric objects can be described by a point process (random point) in a parameter space. These point processes are in general restricted to lie on a manifold in parameter space. For example, consider a line on the plane described by the equation

$$g(\mathbf{x}, \mathbf{p}) = r - x \cos \theta - y \sin \theta = 0 \qquad (2.3)$$

with $\mathbf{x} = [x, y]^T$ and $\mathbf{p} = [r, \theta]^T$ (Figure 2.1). We can describe any 'randomness' of this line as a random point $\mathbf{p}$ in the corresponding parameter space. In general, if the space in which this point is defined is of greater dimension than $\mathbf{p}$ itself, then the process generating this parameter vector will be restricted to a subspace of this defined space. In the case of a line parameterized by $[r, \theta]^T$, we can consider the point

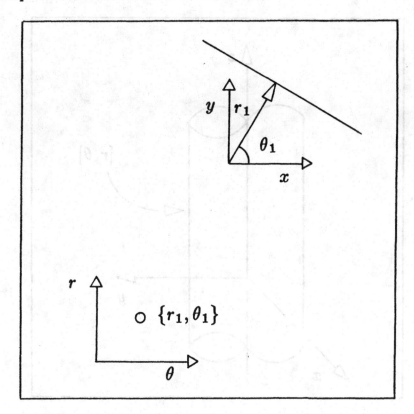

Figure 2.1: A random line on a plane and its corresponding point process in parameter space

process as unconstrained on a plane, or restricted to the surface of an infinite right cylinder (θ orientation and r height) in three dimensions (Figure 2.2), for example. We will generally restrict our interest to spaces in which the point process **p** is unconstrained, although the description in higher dimensional spaces may be useful when two or more *different* geometric objects must be considered simultaneously.

The advantage of describing uncertain geometry by a point in parameter space together with an associated p.d.f., is that to transform uncertain geometries to other coordinate systems, we need only transform their probability distribution func-

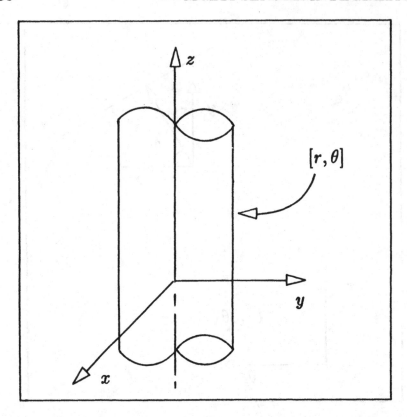

Figure 2.2: The point process of a line restricted to a cylinder.

tions. The representation of all geometric locations and features by point processes lends a homogeneity to the description of stochastic geometries. This considerably simplifies the transformation and manipulation of geometric elements, and provides a powerful framework in which to describe all uncertain geometric features and relations.[2]

---

[2]More recently, Harding [Harding 74] has extended these ideas toward a theory of random sets. In this case $x \subseteq \Re^n$ is the set, and $p \in \Re^m$ a point characterizing this set such that there exists a Borel-measurable function $\Omega(x) \mapsto p$ with the usual closure and uniqueness properties. The point p is considered as a characteristic or indicator of the set x. As such, the theory of random sets is directly applicable to problems in stochastic geometry.

2.4.2 Well-Condition Stochastic Geometries

There are many different functions of the form of Equation 2.1 which can be used to describe a single type of geometric feature. In choosing which representation is most appropriate for a particular feature, we should be concerned with the stability of the parameterization to perturbations or uncertainties. A statistically well-conditioned representation can be of major benefit when features must be manipulated or transformed between coordinate systems. For example, a straight line on a plane could be described by any of the following forms:

$$g(\mathbf{x}, \mathbf{p}) = r - x \cos \theta - y \sin \theta = 0, \qquad \mathbf{p} = [r, \theta]^T$$
$$g(\mathbf{x}, \mathbf{p}) = ax + by + c = 0, \qquad \mathbf{p} = [a, b, c]^T$$
$$g(\mathbf{x}, \mathbf{p}) = y - mx + l = 0, \qquad \mathbf{p} = [m, l]^T$$

The first representation is well defined for all parameter values, although its periodic form can cause confusion. The second representation has the undesirable property that a given feature does not have a unique parameter vector. The third representation cannot parameterize lines parallel to the x-axis.

There are three basic criteria which every representation should satisfy if they are to be considered statistically well-conditioned.

1. Every feature instance must have a unique parameterization. This excludes any representation for which a parameter vector does not exist for all possible feature instances, and it excludes representations where there is a symmetric ambiguity.

2. Every parameterization must be associated with a unique feature. This excludes homogeneous tensor representations or any other parameterization in which the parameter vector is of greater degree than the geometric variety that it describes.

3. The parameter vector must change "smoothly" with physical feature motion. This ensures that small changes in feature do not result in abrupt changes in the describing parameter.

Conditions 1 and 2 require a representation to be a one-to-one (bijective) relation between feature and parameter. Condition 3 is a stability condition (in which it is possible to subsume the first two conditions) which ensures that changes of coordinate system in which the feature is described does not result in rapid changes of the parameter vector.

To analyze this stability condition, we need to consider what happens to the parameter vector $\mathbf{p}$ and it's associated p.d.f $f_g(\mathbf{p})$ when they are moved between coordinate systems. Suppose we start with a representation described by Equation 2.1 and parameterized by the vector $\mathbf{p}$. In general if we transform the point set $\mathbf{x}$ to $\mathbf{x}'$ as $\mathbf{x}' = \mathbf{T}(\mathbf{x})$, the parameter vector describing the set is also changed as $\mathbf{p}' = \mathbf{h}_g(\mathbf{p})$ (say), so that:

$$
\begin{aligned}
\mathbf{g}(\mathbf{x}, \mathbf{p}) &= 0 \\
\text{and} \quad \mathbf{g}(\mathbf{T}(\mathbf{x}), \mathbf{h}_g(\mathbf{p})) &= 0
\end{aligned}
\tag{2.4}
$$

It is important to note that the function $\mathbf{g}$ is the same in any coordinate system, and the transformation $\mathbf{T}(\mathbf{x})$ has the same form for any set $\mathbf{x}$. It follows that when $\mathbf{T}$ is fixed, $\mathbf{h}_g(\mathbf{p})$ is completely defined by the function $\mathbf{g}$.

We want to ensure that for all transformations $\mathbf{T}$ the parameter vector $\mathbf{p}$ and hence the representation $\mathbf{g}$ is well defined. In particular, $\mathbf{p}$ should be well behaved for *differential changes* in coordinate system. Consider the transformation of a parameter vector

$$
\mathbf{p}' = \mathbf{h}_g(\mathbf{p})
\tag{2.5}
$$

from a coordinate system c to c'. Keeping $\mathbf{T}$ fixed, suppose we perturb the vector $\mathbf{p}$ (due to error or noise) by an amount $\delta\mathbf{p}$, so that:

$$
\mathbf{p}' + \delta\mathbf{p}' = \mathbf{h}_g(\mathbf{p} + \delta\mathbf{p})
\tag{2.6}
$$

Using a Taylor series expansion, results in:

$$\delta\mathbf{p}' \approx \nabla_p\mathbf{h}|_p\delta\mathbf{p} \qquad (2.7)$$

The Jacobian $\nabla_p\mathbf{h}$ can be interpreted as a change *magnifier*, describing the change in parameter $\delta\mathbf{p}$ viewed from different coordinate systems. It follows that if $\nabla_p\mathbf{h}$ is well defined as a function of $\mathbf{p}$ then the transformation of a feature representation between arbitrary coordinate systems must also be well behaved.

In statistical terms, $\mathbf{p}$ and $\mathbf{p}'$ are random vectors to which we associate the p.d.f.'s $f_g(\cdot)$ and $f'_g(\cdot)$. We are concerned that these distributions are well behaved when the parameter vectors are transformed between coordinate systems. Consider a differential volume $d\mathbf{p} = dp_1 \cdots dp_m$ of a parameter space. If we interpret a change of coordinate system as a change of variables, and if this transform is bijective, then we can write;

$$dp'_1 dp'_2 \cdots dp'_m = |\nabla_p\mathbf{h}| dp_1 dp_2 \cdots dp_m \qquad (2.8)$$

where $|\nabla_p\mathbf{h}|$ is the determinant of the transform Jacobian. The probability mass associated with a given differential volume must remain constant regardless of the coordinate system in which it is defined, so that:

$$f'_g(\mathbf{p}') dp'_1 dp'_2 \cdots dp'_m = f_g(\mathbf{p}) dp_1 dp_2 \cdots dp_m \qquad (2.9)$$

Substituting Equation 2.8 into this relation gives

$$f'_g(\mathbf{p}')|\nabla_p\mathbf{h}| dp_1 dp_2 \cdots dp_m = f_g(\mathbf{p}) dp_1 dp_2 \cdots dp_m \qquad (2.10)$$

which results in the usual equation for the transformation of probability distributions:

$$f'_g(\mathbf{p}') = \frac{1}{|\nabla_p\mathbf{h}|} f(\mathbf{h}^{-1}[\mathbf{p}']) \qquad (2.11)$$

A detailed development of this equation and its properties can be found in [Papoulis 65].

Consideration of Equation 2.11 shows that if $|\nabla_p \mathbf{h}|$ is well behaved as this vector is transformed between coordinate systems, then the distribution on a parameter vector will be statistically well conditioned. This agrees with the result obtained in Equation 2.7. The analysis of $\nabla_p \mathbf{h}$ can provide an important insight into the effectiveness of a particular geometric representation. Recall that the parameter transform $\mathbf{h}$ can be found uniquely from g by considering arbitrary transforms on the set $\mathbf{x} \subseteq \Re^n$. When $\mathbf{h}$ has been found, the stability of the representation provided by g can be analyzed by considering the Jacobian $\nabla_p \mathbf{h}$ as a function of $\mathbf{p}$.

To illustrate this, we will take the example [Kendall 63] of a straight line on a plane described by either of:

$$g(\mathbf{x}, \mathbf{p}) = ux + vy + 1 = 0, \qquad \mathbf{p} = [u, v]^T$$
$$g(\mathbf{x}, \mathbf{p}) = r - x \cos \theta - y \sin \theta = 0, \qquad \mathbf{p} = [r, \theta]^T$$

We consider arbitrary transforms of the form

$$\begin{bmatrix} x' \\ y' \end{bmatrix} = \begin{bmatrix} \cos \phi & -\sin \phi \\ \sin \phi & \cos \phi \end{bmatrix} \begin{bmatrix} x \\ y \end{bmatrix} + \begin{bmatrix} T_x \\ T_y \end{bmatrix} \qquad (2.12)$$

So that the parameter transforms are:

$$\begin{bmatrix} u' \\ v' \end{bmatrix} = \frac{\begin{bmatrix} u \sin \phi + v \cos \phi \\ u \cos \phi - v \sin \phi \end{bmatrix}}{[1 + T_x(u \cos \phi + v \sin \phi) + T_y(u \sin \phi + v \cos \phi)]} \qquad (2.13)$$

and

$$\begin{bmatrix} r' \\ \theta' \end{bmatrix} = \begin{bmatrix} r + T_x \cos(\theta + \phi) + T_y \sin(\theta + \phi) \\ \theta + \phi \end{bmatrix} \qquad (2.14)$$

These parameter transforms describe the relocation of lines. To find out how stable these transforms are, we need only consider the determinant of the transform Jacobians. Differentiating Equations 2.13 and 2.14 we obtain

$$|\nabla_{u,v} \mathbf{h}| = (u^2 + v^2)^{\frac{3}{2}}$$
$$|\nabla_{r,\theta} \mathbf{h}| = 1$$

Using these in Equation 2.8 interpreted in the form of Equation 2.11 results in the differential measures $(u^2 + v^2)^{-3/2} du\, dv$ and $dr\, d\theta$ These quantities are known as the invariant measure of a representation; the differential probability mass that remains invariant under all rotations and translations appropriate to the Euclidean group. From this example, it is clear that $[r, \theta]^T$ is a natural representation for the line on a plane.[3]

The importance of finding well behaved representations for uncertain geometric objects is that it allows observed features to be manipulated and transformed between coordinate systems in a well defined, robust manner. We have demonstrated a verification procedure for analyzing the statistical properties of a specific representation. Ideally it would be useful to find an algorithm to determine which representation is best for every given feature type. Although we do not know how this can be done, Kendall and Moran [Kendall 62] suggest a more general verification principle which may lead to a solution of this problem.

2.4.3 Stochastic Topology

We are primarily interested in geometric objects embedded in the usual Euclidean space. These objects are conceptually linked together by virtue of the geometric relations resulting from this embedding. It is therefore impossible to consider operations on a geometric object in isolation from other objects in this space. We can describe these links between different objects as a set of relations - a topology - which can be characterized by some simple rules of invariance. These rules can be interpreted as constraints on the set of relations between uncertain geometric objects. For example, consider two uncertain objects related to each other by virtue of their embedding in real space; if the information we have about one object changes,

---

[3]In this example, $dr\, d\theta$ is the *only* measure invariant representation for this geometry [Poincaré 12].

what effect will this have on the second object to which it is related, if we change the relation between these objects, what can be said about the resulting change in object description? This same argument applies to the relations between many different objects, linked together by a network of general relations resulting from their embedding in Euclidean space.

We will call the set of relations on probabilistic geometries a *stochastic topology*. The invariants of this topology will be termed *stochastic invariants*. The invariants of the topology associated with uncertain geometric objects are those properties of the relations between objects that remain unchanged when the parameters **p** that describe the features are "distorted" or perturbed about their nominal values. If we consider the probability distribution function $f_g(\mathbf{p})$ associated with the representation $g(\mathbf{x}, \mathbf{p}) = 0$ as some likelyhood of distortions, then it follows that the stochastic invariants of relations between uncertain geometric objects are exactly those topological invariants associated with the Euclidean set.

The invariant properties of relations between objects provides a very powerful tool with which to maintain consistency between feature descriptions and propagate geometric observations through the robots world model. The reason for this is that the relations between features can be used to provide constraints on possible descriptions. These constraints can be interpreted as information and used, whenever feature descriptions are changed, to apply corresponding changes to related features.

The simplest case of an invariant relation occurs in the loop created by the relations between three geometric objects. In this instance, the vector sum of the relations between objects is zero, regardless of the absolute location of these objects. Consider for example the *random* placement of three points on a plane (Figure 2.3);

$$\mathbf{p}_1 = [x_1, y_1]^T, \quad \mathbf{p}_2 = [x_2, y_2]^T, \quad \mathbf{p}_3 = [x_3, y_3]^T \qquad (2.15)$$

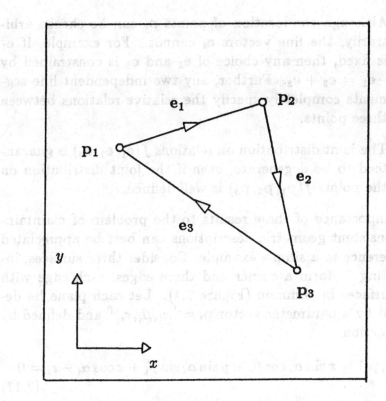

Figure 2.3: The general three-node three-arc network.

and the (random) relations created between them:

$$e_1 = p_2 - p_1, \quad e_2 = p_3 - p_2, \quad e_3 = p_1 - p_3, \qquad (2.16)$$

There are three important observations that can be made about this simple case:

1. Regardless of the initial distribution on the placement of points, the sum of the relations between these points will be zero with probability 1^4, so that $e_1 + e_2 + e_3 = 0$.

---

[4]Providing the distribution on points is not degenerate, the event that the points are co-linear has measure zero. However this is *not* the impossible event.

2. Although the location of points p_i can be chosen arbitrarily, the line vectors e_i cannot. For example, if e_1 is fixed, then any choice of e_2 and e_3 is constrained by $-e_1 = e_2 + e_3$. Further, any two independent line segments completely specify the relative relations between three points.

3. The joint distribution on relations $f(e_1, e_2, e_3)$ is guaranteed to be degenerate, even if the joint distribution on the points $f(p_1, p_2, p_3)$ is well defined.

The importance of these results to the problem of maintaining consistent geometric descriptions can best be appreciated by reference to a simple example. Consider three surfaces, intersecting to form a corner and three edges, each edge with two surfaces in common (Figure 2.4). Let each plane be described by a parameter vector $p_i = [\alpha_i, \beta_i, r_i]^T$ and defined by the function

$$g(\mathbf{x}, p_i) = x \sin \alpha_i \cos \beta_i + y \sin \alpha_i \sin \beta_i + z \cos \alpha_i - r_i = 0$$

$$(2.17)$$

with $\mathbf{x} = [x, y, z]^T$. These three planes can be represented as points p_1, p_2, p_3 in the parameter space $[\alpha, \beta, r]^T$ (Figure 2.5). The location of these points can be chosen arbitrarily and still be interpreted as a consistent geometric objects with three edges meeting at a common corner. The location of the corner can be uniquely determined by solving the three simultaneous equations (Equation 2.17) for $\mathbf{x} = [x, y, z]^T$. A corner in Euclidean space corresponds to the surface generated by three points in parameter space. Any one edge can be found by solving for x and y in terms of z between any two corresponding equations of the planes. An edge in Euclidean space corresponds to a line generated by two points in parameter space.

The network of points in parameter space is equivalent to the network in Euclidean space and consequently shares the

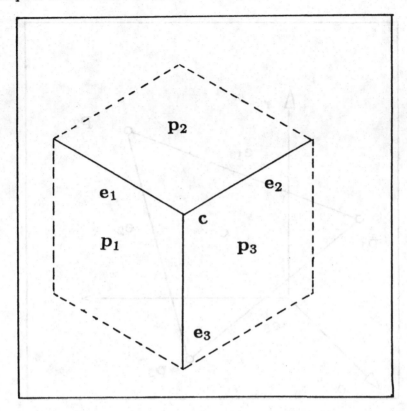

Figure 2.4: Three planes intersecting to form three edges and a corner.

same properties: If any one plane description p_i is changed to some new value p_i', a consistent geometric object, with edges and a corner, is maintained without the need to change any other plane description. Conversely, if any one edge description e_i is changed, both of the other edge descriptions must also be changed if consistency is to be maintained. Figure 2.5 illuminates this, showing that the change in edge descriptions must be constrained to satisfy, in some form, a relation $e_2' + e_3' = -e_1'$.

As before we can consider these changes as a measure of the uncertainty in the description of geometric features. In

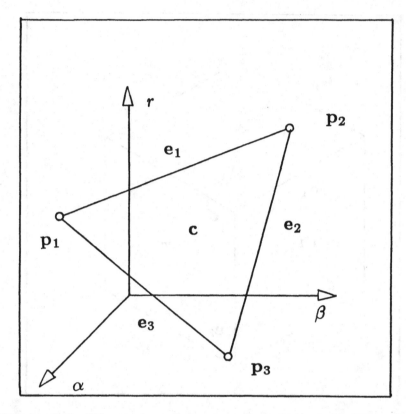

Figure 2.5: The parameter space for planes, showing three planes, three edges and a corner.

this case any change in feature estimate or distribution of uncertainty is constrained by it's relation to other descriptions. Consider again the example of three surfaces forming three edges and a corner. Suppose we initially have an estimate of the object parameters with some measure of uncertainty, and suppose we now make a new observation of an edge; e_i. Combining prior and observed estimates of the edge parameters results in some new estimate e'_i. To maintain geometric consistency of the object, this new information must be propagated through the constraining relations to provide new consistent estimates for the other edges of the object. This idea of consistency, constraint and propagation is fundamental to the representation of uncertain geometry and the interpretation of sensory data.

We can generalize these simple principles to any number of different features and relations: Consider a connected, directed network consisting of n nodes representing features or object locations and m arcs $(n \leq m)$ representing geometric relations between nodes, together forming r loops. Index the nodes (features) by p_i $(i = 1, \cdots, n)$ and label the arcs with random variables e_j, $(j = 1, \cdots, m)$ representing the *uncertain* relations between features. This network can be described by either a path matrix C $(r \times m)$ or an incidence matrix M $(n \times m)$. The path matrix describes the arcs contained in each network loop. The elements of C can only take the values 0 or 1 describing if an arc is in a loop. The path matrix need only contain a basis set of independent loops to completely define the network. The incidence matrix describes which nodes are connected by each arc. The elements of M can only take on the values 0, 1 or -1 denoting incidence of an arc out of or into a node. For example the three node three arc network of Figure 2.3 has:

$$M = \begin{bmatrix} 1 & 0 & -1 \\ -1 & 1 & 0 \\ 0 & -1 & 1 \end{bmatrix} \quad \text{and} \quad C = \begin{bmatrix} 1 & 1 & 1 \end{bmatrix} \qquad (2.18)$$

It can be shown [Chartrand 68] that for all directed connected networks:

$$CM^T = 0 \quad \text{or} \quad MC^T = 0 \qquad (2.19)$$

This means that for every node contained in a loop, there are two arcs, one into and one out of the node, which are part of this loop.

Let $e = [e_1, \cdots, e_m]^T$ be the random vector of arc labelings and define $\gamma = [\gamma_1, \cdots, \gamma_r]^T$ to be the random vector of "distances" around a loop, so that

$$Ce = \gamma \qquad (2.20)$$

describes the structure of the network. A connected directed network is such that if we introduce another node $n+1$ to the network and connect it to all other nodes with n new arcs labeled with random variables x_i, $(i = 1, \cdots, n)$, then $\mathbf{x} = [x_1, \cdots, x_n]^T$ must satisfy

$$\mathbf{e} = \mathbf{M}^T \mathbf{x} \qquad (2.21)$$

The following results are true for arbitrary probability distributions on arc labels $\mathbf{e}$:

Result 2.1: If $\mathbf{e}$ is the vector of random variables labeling a constrained network then $\mathbf{e}$ is in the null-space of $\mathbf{C}$ with probability 1:

$$\mathbf{Ce} = \mathbf{0} \qquad (2.22)$$

Proof: We have $\mathbf{Ce} = \gamma$ and $\mathbf{e} = \mathbf{M}^T \mathbf{x}$ with $\mathbf{x}$ an arbitrary n dimensional random vector. Then $\mathbf{CM}^T \mathbf{x} = \gamma$, but $\mathbf{CM}^T = \mathbf{0}$ so $\gamma = 0$ with probability 1.

Result 2.2: The variance-covariance matrix $\mathbf{\Lambda}_e$ of the vector $\mathbf{e}$ is singular.
Proof: $\mathbf{Ce} = \mathbf{0}$ so with $\hat{\mathbf{e}} = E[\mathbf{e}]$, [5]

$$\mathbf{C}E\left[(\mathbf{e} - \hat{\mathbf{e}})(\mathbf{e} - \hat{\mathbf{e}})^T\right]\mathbf{C}^T = \mathbf{C}\mathbf{\Lambda}_e\mathbf{C}^T = \mathbf{0} \qquad (2.23)$$

and as $\mathbf{\Lambda}_e$ is symmetric, it must also be singular.

Result 2.3: For any symmetric matrix $\mathbf{\Lambda}_x$, the matrix $\mathbf{M}^T\mathbf{\Lambda}_x\mathbf{M}$ is singular.
Proof: $\mathbf{e} = \mathbf{M}^T\mathbf{x}$ so $\mathbf{\Lambda}_e = \mathbf{M}^T\mathbf{\Lambda}_x\mathbf{M}$ is singular. As $\mathbf{x}$ is arbitrary, so is $\mathbf{\Lambda}_x$.

These results can be applied to relations between all types of locations and geometric features. They provide a mechanism

---

[5] E is the expectation operator; $E[x] = \int_X x f(x) dx$

with which consistency between descriptions can be maintained
and by which relations between objects can be described. This
consistency requirement also results in geometric constraints
through which changes in description or new information can
be propagated[6]. These three results will be used in later chapters to develop consistent sensor communication and integration methods.

2.5 Manipulating Geometric Uncertainty

It is important to consider how information about geometric
objects and features change when they are described in different coordinate systems. This becomes especially apparent
when we have mobile sensor systems that change their perspective of the environment. In addition, it is often important
to be able to represent information about one geometric object
in terms of another type of object. This occurs for example if
we wish to find the contribution of an edge or curvature to a
surface description or when integrating information from sensors that provide disparate geometric observations. We shall
generally consider the case of a geometric feature described by
Equation 2.1 with $\mathbf{x} \in \Re^6$ and $\mathbf{p} \in \Re^m$. To each uncertain geometric object defined by g and parameterized by $\mathbf{p}$ we associate
a distribution function $f_g(\mathbf{p})$ which describes the likelyhood of
a particular feature instance. We will require that $f_g(\mathbf{p})$ is not
degenerate; where the probability mass is concentrated on a
hypersurface in the parameter space, or where perfect information exists on one or more degrees of freedom. We do not
preclude the case where the distribution function is improper,
where there is infinite uncertainty or no information in one or

---

[6]It should be noted that this development of stochastic networks has
much in common with other network results, and is particularly applicable
in the field of stochastic automata and analogue networks [Poggio 85].

more degrees of freedom. In all cases, the density $f_g(\mathbf{p})$ will be assumed to have finite moments of all orders.

There are two basic operations that we will want to perform on uncertain geometric descriptions: Changing the coordinate system in which the object is described; corresponding to transforming the feature parameter vector. Changing the description of the object itself; corresponding to both a transformation of parameter vector and a change of representing function. Both of these operations require that the parameter density function be changed in addition to any purely geometric change. This is equivalent to a transformation of probability distributions.

2.5.1   Transforming Probability

The key to manipulating geometric uncertainty is to be able to transform the information or probability density function on a feature available in one form in to another form of interest. This is the case when comparing surfaces (for planning and mating operations) or viewing objects from different locations. Consider first the transformation of a feature between coordinate systems. Suppose the random vector $\mathbf{p}_i$ describes a geometric feature with respect to a coordinate system i. In this frame $\mathbf{p}_i$ has probability density function $f^i(\cdot)$. If we describe this feature with respect to another coordinate frame j as $\mathbf{p}_j$ then in general the relation between the two density functions $f^j(\cdot)$ and $f^i(\cdot)$ will be described by Equation 2.11. In general if the transformation $\mathbf{p}_j = {}^j\mathbf{h}_i(\mathbf{p}_i)$ is non-linear, the calculation of the transformed distribution functions using Equation 2.11 can be very complex. Changing between feature descriptions increases this complexity: In this case $\mathbf{p}_i \in \Re^m$ and $\mathbf{p}_j \in \Re^l$ are usually of different dimension so that the transform ${}^j\mathbf{h}_i \colon \Re^m \to \Re^l$ is not always well defined in a deterministic sense. If $m > l$, when features are aggregated for example, then $f^j(\cdot)$ will be degenerate (over-constrained). Conversely if $m < l$, when higher order descriptions are calculated, then $f^j(\cdot)$ will be improper (under-constrained). However, even if the transform is not well defined in a deterministic sense, it can always be properly represented in a stochastic sense. Where ^jh_i is over-constrained, the dimension of p_j can be reduced to satisfy the topological properties of the feature space. If ^jh_i is under-constrained, indeterminacy can be accounted for by providing no information (infinite uncertainty) in one or more degrees of freedom.

The geometric complexity of the transform ^jh_i can make the solution of Equation 2.11 very difficult. This situation is further aggravated] when sequences of transformations $p_i \rightarrow p_j \rightarrow {}^kp$ are considered. This problem arises in even very simple situations: Brooks [Brooks 84b,85] has described a method of constructing uncertainty manifolds generated by the motion of a mobile robot in a plane. The robot moves with uniformly distributed uncertainty in location and heading. After each (stochastic) motion the robot arrives at a location bounded by a closed manifold in $\Re^3$. This has two main problems, firstly the continuing assumption of uniformity in the location distribution after each step is invalidated by Equation 2.11, (although the exact density is not hard to calculate). Secondly after each motion the manifold gets increasingly difficult to describe as a geometric surface, indeed after several motions, Brooks (justifiably) approximates the manifold by a cylinder.

There are two parts to this complexity issue: the geometry associated with the transformations and the underlying process described by the distribution functions f. There is unfortunately little that we can do to alleviate the geometric complexity of the environment. Even in the simplest case, where a change in feature description is linear with respect to Euclidean transformations, we may often have to choose another description to satisfy the requirements of representation stability. The transformation of probability distributions faces many of the same problems as do geometric transforms. Ideally we should find a distribution function f_g which has invariant measure over all objects and transforms of the Euclidean group; so that $f_g(\cdot)$ is the same function in all coordinate systems. In general such a distribution function does not exist, however it can often be found for particular objects or motions. We can relax the invariance requirement and just look for a density which is *conjugate* over all possible translations and rotations. That is, after transforming the random variable by $\mathbf{p}_j = {}^j\mathbf{h}_i(\mathbf{p}_i)$, the two distributions f^j and f^i on $\mathbf{p}_j$ and $\mathbf{p}_i$ come from the same family $\mathcal{F}$ of distributions: $f^j, f^i \in \mathcal{F}$, which can be parameterized by some simple transformation ($\mathcal{F}$ is conjugate over all transforms ${}^j\mathbf{h}_i$). Given the possible generality of the transforms ${}^j\mathbf{h}_i$ it is clear that such a family would be difficult to find.

In general, the only way around these complexity problems is to use an approximation to the exact transformation of information.

2.5.2 Approximate Transforms

An alternative to calculating and propagating exact probability distributions is to expand the transformation equation in to its corresponding moments and truncate this series at some appropriate point. Consider the transformation

$$\mathbf{p}_j = {}^j\mathbf{h}_i(\mathbf{p}_i) \tag{2.24}$$

As expectation is a linear operator, we have:

$$\hat{\mathbf{p}}_j = E[\mathbf{p}_j] = E[{}^j\mathbf{h}_i(\mathbf{p}_i)] = {}^j\mathbf{h}_i(E[\mathbf{p}_i]) = {}^j\mathbf{h}_i(\hat{\mathbf{p}}_i) \tag{2.25}$$

Taking perturbations about the distribution mean;

$$\hat{\mathbf{p}}_j + \delta\mathbf{p}_j = {}^j\mathbf{h}_i(\hat{\mathbf{p}}_i + \delta\mathbf{p}_i) \tag{2.26}$$

and expanding using a Taylor series,

$$\hat{\mathbf{p}}_j + \delta\mathbf{p}_j = {}^j\mathbf{h}_i(\hat{\mathbf{p}}_i) + \left(\frac{\partial {}^j\mathbf{h}_i}{\partial \mathbf{p}_i}\right)\bigg|_{p=\hat{p}} \delta^i\mathbf{p} + O(\mathbf{h}^2) \tag{2.27}$$

results in the first order relation:

$$\delta \mathbf{p}_j \approx \left(\frac{\partial^j \mathbf{h}_i}{\partial \mathbf{p}_i}\right)\bigg|_{\mathbf{p}=\hat{\mathbf{p}}} \delta \mathbf{p}_i \qquad (2.28)$$

Squaring Equation 2.28 and taking expectations

$$
\begin{aligned}
{}^j\mathbf{\Lambda}_p &= E\left[\delta \mathbf{p}_j \delta \mathbf{p}_j^T\right] \\
&\approx E\left[\left(\frac{\partial^j \mathbf{h}_i}{\partial \mathbf{p}_i}\right)\delta \mathbf{p}_i \delta \mathbf{p}_i^T \left(\frac{\partial^j \mathbf{h}_i}{\partial \mathbf{p}_i}\right)^T\right] \\
& \qquad\qquad\qquad\qquad\qquad\qquad (2.29) \\
&= \left(\frac{\partial^j \mathbf{h}_i}{\partial \mathbf{p}_i}\right)E\left[\delta \mathbf{p}_i \delta \mathbf{p}_i^T\right]\left(\frac{\partial^j \mathbf{h}_i}{\partial \mathbf{p}_i}\right)^T \\
&= \left(\frac{\partial^j \mathbf{h}_i}{\partial \mathbf{p}_i}\right){}^i\mathbf{\Lambda}_p\left(\frac{\partial^j \mathbf{h}_i}{\partial \mathbf{p}_i}\right)^T
\end{aligned}
$$

Approximating Equation 2.11 by Equations 2.25 and 2.29 is equivalent to a linearization of Equation 2.11. The mean values of the transformation follow the usual laws of geometry (Equation 2.25), and the distribution variances are transformed as differential changes (Equation 2.29). Providing the density function $f^i(\cdot)$ has finite moments of all orders, we could, in theory, calculate the transformed density by an infinite expansion of transformed moments. These successive moments can be found by equating higher order terms from Equation 2.27. In practise, using higher order terms is neither viable nor desirable, as any computational simplicity will be lost. However, if this linear approximation is to be used, we should first attempt to justify its validity.

2.5.3 Properties of Transformations

The mean values of the parameter distribution function are transformed in exactly the same manner as in the case of a purely deterministic environment model (Equation 2.25). The first moment (variance) of this distribution is transformed by pre and post multiplying by the transform Jacobian (Equation 2.29). If the matrix $\frac{\partial h}{\partial p}$ becomes singular or ill-conditioned at some value of p then this first order approximation will be invalidated. However, in our development of conditions for the stability of uncertain feature representations, we required that the transform Jacobian be well-behaved under all transformations in order that the feature description be statistically well-conditioned. Thus the fact that a geometric representation is stochastically well conditioned ensures that the linearized transformation is a good approximation to the true distribution.

In linearizing the transformation of a probability distribution we must make an implicit assumption that both the initial density $f(\cdot)$ and the transformed density $f'(\cdot)$ are symmetric, unimodal and have finite moments of all orders. Specifically, if $f(\cdot)$ satisfies these conditions, we must be able to guarantee that the same applies to $f'(\cdot)$ under all possible transformations of p. In the case where the transformation corresponds to a change in coordinate system, this can easily be satisfied. However, in the case corresponding to a change in representation, care must be taken to check that the resulting distribution meets these conditions. For example, consider the line formed by the intersection of two infinite planes in three dimensions[7]. If the distribution function defined on the parameter vector of each plane is symmetric and unimodal, what can be said about the distribution on the parameter vector of the resulting line? In theory, each such case must be considered individually; by checking that the transformed distribution is unimodal and symmetric, with density tails enclosed by a Gaussian. In practice however, the stability requirement ensures that any valid parameter transformation is locally smooth, so that these conditions are usually satisfied.

---

[7]The correspondence problem in a line-based stereo algorithm [Ayache 86]

The linear approximation to a transformed density and its resulting conditions on the feature representation, lead inexorably to an assumption of Gaussianity. We shall make a case for considering the probability distribution on the parameter vector describing a feature to be jointly Gaussian. A detailed technical development of these arguments may be found in [Berger 85, Durrant-Whyte 86b]. It is well known that a jointly normal distribution is conjugate with respect to addition of random vectors. This results in very simple algorithms for the aggregation of information. The central limit theorem states that the continued aggregation of any independent identically distributed random variables will converge to a Gaussian density. Aggregation of information in this way results in convergence to a Gaussian decision rule regardless of initial distribution. This effect is called the principle of stable estimation. Initial robustness can be obtained by forcing variances to be large enough to enclose all possible distributions. This process can result in information loss, and care must be taken to ensure suitable distribution-tail behavior is obtained. If we assume Gaussianity, then the mean $\hat{p}$ and variance Λ_p are sufficient information to completely define the feature density function. This means that Equations 2.25 and 2.29 are all we need to transform information amongst coordinate systems. The following well known results are true for jointly Gaussian vectors $\mathbf{x}$.

Result 2.4: If $\mathbf{x}$ is an n dimensional jointly normal random vector with mean $\hat{\mathbf{x}}$ and variance-covariance Λ_z then with $\mathbf{y} = \mathbf{A}\mathbf{x}$ we have [Bickell 77]:

1. $\mathbf{y}$ is jointly normal with mean $\hat{\mathbf{y}} = \mathbf{A}\hat{\mathbf{x}}$ and variance $\Lambda_y = \mathbf{A}\Lambda_z\mathbf{A}^T$

2. Λ_y will be singular if $\mathbf{A}\mathbf{A}^T$ is singular and only if $\mathbf{A}\Lambda_z\mathbf{A}^T$ is singular.

3. If $\mathbf{A}$ has rank n and $\mathbf{\Lambda}_x$ is non-singular, then $\mathbf{\Lambda}_x^{-1} = \mathbf{A}^T \mathbf{\Lambda}_y^{-1} \mathbf{A}$

The ease with which Gaussian distributions can be manipulated allows the development of fast and efficient algorithms for the manipulation of uncertain geometry.

2.6   Gaussian Geometry

We maintain that a reasonable policy for modeling and propagating geometric uncertainty will always tend to Gaussianity and in case, can be approximated by such. We will present some elements of a theory of Gaussian geometry, where all geometric objects and features are parameterized by a jointly Gaussian random vector. We shall show how information and uncertainty can be manipulated in an efficient manner, providing a mechanism for reasoning about the effects of geometric uncertainty. We will first consider the problem of changing coordinate systems, and then apply these ideas to a change in feature description.

2.6.1   Changing Locations

It is important to consider how information about geometric objects and features change when they are described in different coordinate systems. This becomes especially apparent when we have mobile sensor systems that change their perspective of the environment information. Our motivating example will be the case of a six dimensional location vector composed of a position and a roll-pitch-yaw orientation. The case for specific features can often be reduced to a problem of locating coordinate systems by attaching a reference frame to the feature of interest. Considering a feature transform as a coordinate transform usually results in an underconstrained change in representation due to the ambiguity involved in attaching frame to feature. The direct manipulation of feature representations follows a similar but usually simpler formulation.

Consider the transformation of a random vector

$$\mathbf{p} = [x, y, z, \phi, \theta, \psi]^T$$

that parameterizes an oriented point in $\Re^3$. The transformation of the parameter mean $\hat{\mathbf{p}}$ follows the usual laws of geometry (Equation 2.25). Of more interest is the information content of representations in different coordinate frames as described by the transformation of variance-covariance matrix $\mathbf{\Lambda}_p$ or the information matrix $\mathbf{\Lambda}_p^{-1}$. Suppose the function $^j\mathbf{h}_i$ that transforms a description vector $\mathbf{p}_i$ to $\mathbf{p}_j$ can be described by a homogeneous transform $^j\mathbf{T}_i$ that transforms a coordinate frame i to a frame j. The mean $\hat{\mathbf{p}}_i$ in frame i is transformed to $\hat{\mathbf{p}}_j$ in frame j by Equation 2.25, the variance-covariance matrix $^i\mathbf{\Lambda}_p$ in frame i is transformed to $^j\mathbf{\Lambda}_p$ in frame j by the transformation Jacobian $^j\mathbf{J}_i = \frac{\partial \mathbf{h}}{\partial \mathbf{p}}$:

$$^j\mathbf{\Lambda}_p = {}^j\mathbf{J}_i{}^i\mathbf{\Lambda}_p{}^j\mathbf{J}_i^T \tag{2.30}$$

If

$$^j\mathbf{T}_i = \begin{bmatrix} \mathbf{n} & \mathbf{o} & \mathbf{a} & \mathbf{q} \\ 0 & 0 & 0 & 1 \end{bmatrix}$$

then the Jacobian $^j\mathbf{J}_i$ can be found from the transformation [Paul 81] as:

$$\mathbf{J} = \begin{bmatrix} \mathbf{r} & \mathbf{m} \\ 0 & \mathbf{r} \end{bmatrix}, \quad \mathbf{r} = \begin{bmatrix} \mathbf{n}^T \\ \mathbf{o}^T \\ \mathbf{a}^T \end{bmatrix}, \quad \mathbf{m} = \begin{bmatrix} (\mathbf{q} \times \mathbf{n})^T \\ (\mathbf{q} \times \mathbf{o})^T \\ (\mathbf{q} \times \mathbf{a})^T \end{bmatrix} \tag{2.31}$$

where $\mathbf{r}$ is the 3×3 rotation matrix and $\mathbf{m}$ the 3×3 magnification matrix.

We state some results without proof:

Result 2.5:

1. $\mathbf{rr}^T = \mathbf{I}$: $\mathbf{r}$ is Hermatian.

2. mr^T is skew symmetric;

$$mr^T = \begin{bmatrix} 0 & q \cdot a & -q \cdot o \\ -q \cdot a & 0 & q \cdot n \\ q \cdot o & -q \cdot n & 0 \end{bmatrix} \qquad (2.32)$$

hence $rm^T = -mr^T$, or; $m^T = -r^T mr^T$

3. mm^T is symmetric

4. $mm^T = rmm^T r^T = qq^T + rqq^T r^T$; mm^T is rotation invariant.

Using these results, the Jacobian, it's transpose and inverse can be found from T:

$$J_T = \begin{bmatrix} r & m \\ 0 & r \end{bmatrix} \qquad J_T^T = \begin{bmatrix} r^T & 0 \\ m^T & r^T \end{bmatrix} \qquad (2.33)$$

As $J_T^{-1} = J_{T^{-1}}$, we have:

$$J_T^{-1} = J_{T^{-1}} = \begin{bmatrix} r^T & m^T \\ 0 & r^T \end{bmatrix}, \quad J_T^{-T} = J_{T^{-1}}^T = \begin{bmatrix} r & 0 \\ m & r \end{bmatrix} \quad (2.34)$$

With T the transform between coordinate frames i and j, from Equation 2.31 and Equations 2.33 and 2.34, we have the following relations:
Variance:

$$^j\Lambda_p = J_T{}^i\Lambda_p J_T^T, \qquad ^i\Lambda_p = J_{T^{-1}}{}^j\Lambda_p J_{T^{-1}}^T = J_T^{-1j}\Lambda_p J_T^{-T} \quad (2.35)$$

Information:

$$^j\Lambda_p^{-1} = J_T^{-Ti}\Lambda_p^{-1}J_T^{-1} \qquad ^i\Lambda_p^{-1} = J_T^{Tj}\Lambda_p^{-1}J_T \qquad (2.36)$$

We will now consider the effect of moving information among coordinate frames by the transformation of the covariance and information matrices by the appropriate Jacobian. Let

$$^i\Lambda_p = \begin{bmatrix} ^i\Lambda_{11} & ^i\Lambda_{12} \\ ^i\Lambda_{12}^T & ^i\Lambda_{22} \end{bmatrix}$$

be the variance-covariance of **p** described by appropriate 3×3 variance matrices for position, orientation and their cross correlation. Neglecting cross correlations ($^i\Lambda_{12} = 0$), we obtain:

$$^j\Lambda_p = \mathbf{J}_T{}^i\Lambda_p\mathbf{J}_T^T = \begin{bmatrix} \mathbf{r}^i\Lambda_{11}\mathbf{r}^T + \mathbf{m}^i\Lambda_{22}\mathbf{m}^T & \mathbf{m}^i\Lambda_{22}\mathbf{r}^T \\ \mathbf{r}^i\Lambda_{22}\mathbf{m}^T & \mathbf{r}^i\Lambda_{22}\mathbf{r}^T \end{bmatrix} \quad (2.37)$$

The transformed position variance $^j\Lambda_{11} = \mathbf{r}^i\Lambda_{11}\mathbf{r}^T + \mathbf{m}^i\Lambda_{22}\mathbf{m}^T$ has an interesting interpretation. Clearly $\mathbf{r}^i\Lambda_{11}\mathbf{r}^T$ is just the position variance rotated to the new perspective, the additional term $\mathbf{m}^i\Lambda_{22}\mathbf{m}^T$ is the magnification of orientation uncertainties due to the distance between coordinate frames. Figure 2.6 shows that as $|\mathbf{q}|$ increases in size, so does position uncertainty, much in the same way that small changes in a cameras pan-tilt angle results in large changes of viewpoint. The terms $\mathbf{r}^i\Lambda_{22}\mathbf{m}^T$ reflect this correlation (relation) between the uncertainty in position and orientation.

Often we only have partial information on the vectors **p** (infinite uncertainty in one or more degrees of freedom), whence the distribution function $f_p(\cdot)$ is improper and Λ is not strictly defined. In this case (and in general) it is easier and safer to work with the information matrix $\Sigma = \Lambda^{-1}$, where the diagonal zero elements of Σ imply that no information is available in the associated degree of freedom (infinite uncertainty). Following similar arguments to those above, let:

$$^i\Sigma_p = \begin{bmatrix} ^i\Sigma_{11} & ^i\Sigma_{12} \\ ^i\Sigma_{12}^T & ^i\Sigma_{22} \end{bmatrix}$$

Neglecting cross information ($^i\Sigma_{12} = 0$), (in this case we will also obtain $\Sigma_{11} = \Lambda_{11}^{-1}, \Sigma_{22} = \Lambda_{22}^{-1}$) we have:

$$^j\Sigma_p = \mathbf{J}_{T^{-1}}^T{}^i\Sigma_p\mathbf{J}^{T^{-1}} = \begin{bmatrix} \mathbf{r}^i\Sigma_{11}\mathbf{r}^T & \mathbf{r}^i\Sigma_{11}\mathbf{m}^T \\ \mathbf{m}^i\Sigma_{11}\mathbf{r}^T & \mathbf{m}^i\Sigma_{11}\mathbf{m}^T + \mathbf{r}^i\Sigma_{22}\mathbf{r}^T \end{bmatrix}$$

$$(2.38)$$

The term $^j\Sigma_{22} = \mathbf{m}^i\Sigma_{11}\mathbf{m}^T + \mathbf{r}^i\Sigma_{22}\mathbf{r}^T$ is the dual of the term $^j\Lambda_{11}$ of Equation 2.37 and has an important but less obvious

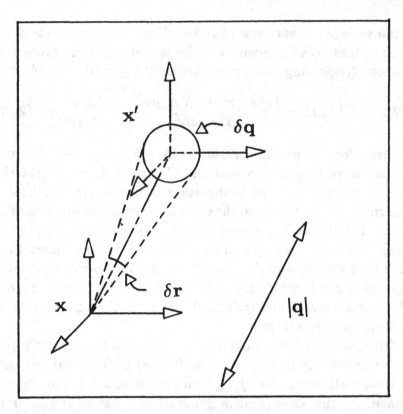

Figure 2.6: Schematic diagram of the magnification of position errors due to orientation errors following a change in location

interpretation. Again $r^i \Sigma_{22} r^T$ is just the rotated orientation information, the additional component $m^i \Sigma_{11} m^T$ is the *extra* information gained from the position information. Intuitively, the more information we have about position (the larger the elements of Σ_{11}), or the further our new perspective is away from the original coordinate frame, the better the orientation information we obtain. As before, the terms $m^i \Sigma_{11} r^T$ reflect this correlation between orientation and position information.

Equation 2.38 has a number of important consequences: It enables information in one coordinate system to be described in another frame of reference which in turn provides an ability to reason about the effects of uncertainty on geometric descriptions of a robot environment. The effect of partial information, or conversely very accurate information, about geometric observations becomes clear by noting the transformation of the elements of the matrices Σ_{11} and Σ_{22}.

This ability to transform uncertainty between coordinate systems also allows sensor observations obtained in one coordinate system to be transformed to another sensor's coordinate system so that information may be integrated. Correspondingly, if a sensor is mobile, then these transformations can be used to develop sensor strategies, making decisions as to where a sensor should be placed to best observe a given feature [Durrant-Whyte 87a].

2.6.2   Changing Feature Descriptions

It is often important to be able to change the way in which a feature is described; to provide other levels of geometric representation or to allow disparate descriptions to be compared. This occurs, for example, if we wish to make inferences about one type of geometric feature when we only have information about another different feature, or if we wish to combine sensor observations made of many different types of geometric features. Transforming information from one description to another involves not only a transformation of information, but also supplying an interpretation that accounts for a change in constraints.

Let $g_i(x, p_i) = 0$ and $g_j(x, p_j) = 0$ describe two physically disparate geometric features with $p_i \in \Re^l$ and $p_j \in \Re^m$ defined on $x \in \Re^n$. Suppose these two features are related so that the (geometric) function ${}^j h_i : \Re^l \rightarrow \Re^m$ maps the parameter vector of one feature to that of the other $p_j = {}^j h_i(p_i)$. In general the functions ${}^j h_i$ are not well defined in a *deterministic* sense. This is because the two related features may have differing degrees of freedom as described by the different dimension of their associated parameter vectors. For example, the observation of an edge by a sensor implies the existence of two intersecting surfaces. Given the equations describing these surfaces, we could, in theory, derive the equation representing the resulting edge. Conversely, given the equation of the observed edge, the equation of either surface can not be derived without using some other constraining information. However, if the functions $^j\mathbf{h}_i$ are considered in a *stochastic* sense then this lack of constraint can be represented as partial information and described by placing zero elements in the information matrix corresponding to the indeterminate degrees of freedom of the transformation. In the edge to surface example, the equations of the edge and some *arbitrary* surface can be used to calculate a resultant second surface. The parameters of this arbitrary surface can then be assigned zero information, thus providing a stochastic description of possible surfaces, resulting from a single edge equation, in terms of a set of free parameters.

Suppose we initially have a feature $\mathbf{g}_i$ parameterized by $\mathbf{p}_i$ which we wish to convert to a new feature description $\mathbf{g}_j$ parameterized by $\mathbf{p}_j = {}^j\mathbf{h}_i(\mathbf{p}_i)$. If our information about the feature parameter vector $\mathbf{p}_i$ is described by a Gaussian probability distribution as $\mathbf{p}_i \sim N(\hat{\mathbf{p}}_i, \mathbf{\Lambda}_i)$, then we can approximate the information available to describe this new parameter vector by another Gaussian distribution;

$$\mathbf{p}_j \sim N(\hat{\mathbf{p}}_j, \mathbf{\Lambda}_j) \tag{2.39}$$

with

$$\hat{\mathbf{p}}_j = {}^j\mathbf{h}_i(\hat{\mathbf{p}}_i) \tag{2.40}$$

and

$$\mathbf{\Lambda}_j = \left(\frac{\partial^j\mathbf{h}_i}{\partial\mathbf{p}_i}\right)\mathbf{\Lambda}_i\left(\frac{\partial^j\mathbf{h}_i}{\partial\mathbf{p}_i}\right)^T \tag{2.41}$$

or

$$\mathbf{\Lambda}_j^{-1} = \left(\frac{\partial^j\mathbf{h}_i^{-1}}{\partial\mathbf{p}_i}\right)\mathbf{\Lambda}_i^{-1}\left(\frac{\partial^j\mathbf{h}_i^{-1}}{\partial\mathbf{p}_i}\right)^T \tag{2.42}$$

The fact that $\mathbf{p}_j$ and $\mathbf{p}_i$ are of different dimension is accounted for by the dimension of the Jacobian. If the transformation is over-constrained, then the resulting Λ_j will be singular (Result 2.4). If the transformation is under-constrained then Λ_j^{-1} must be calculated, to provide for infinite uncertainty in the unconstrained degrees of freedom.

This result can easily be extended to consider the aggregation of feature descriptions by defining $\mathbf{p}_i = [\mathbf{p}_{i,1}^T, \cdots, \mathbf{p}_{i,k}^T]^T$ with $\mathbf{p}_{i,l} \sim N(\hat{\mathbf{p}}_{i,l}, \Lambda_{i,l})$ and letting $\mathbf{p}_j = {}^j\mathbf{h}_i(\mathbf{p}_{i,1}, \cdots, \mathbf{p}_{i,k})$. If the $\mathbf{p}_{i,l}$ are statistically independent, this results in an information transformation of the form

$$\hat{\mathbf{p}}_j = {}^j\mathbf{h}_i(\hat{\mathbf{p}}_{i,1}, \cdots, \hat{\mathbf{p}}_{i,k})$$

$$\Lambda_j = \sum_{l=1}^{k} \left(\frac{\partial^j \mathbf{h}_i}{\partial \mathbf{p}_{i,l}}\right) \Lambda_{i,l} \left(\frac{\partial^j \mathbf{h}_i}{\partial \mathbf{p}_{i,l}}\right)^T$$

To illustrate this transformation process we will consider the simple example of matching infinite planes in a line-based stereo correspondence problem[8]. Suppose we have two cameras (a stereo pair) each observing two-dimensional line segments in image space (Figure 2.7). From these two sets of line observations, we wish to construct a single three-dimensional line. This problem can be approached in two ways, either by assuming each observed two-dimensional line segment is already a three dimensional edge with infinite uncertainty perpendicular to the image plane, or by representing the line segments in each image as infinite planes whose intersection will form an edge. The former case has the advantage that the line segments in one image can be used to generate line-hypotheses in the other image plane by applying the transformations developed in Section 2.6.1. The latter case is a more illustrative example of the techniques described in this section.

---

[8]This example was suggested by Faugeras [Faugeras 86].

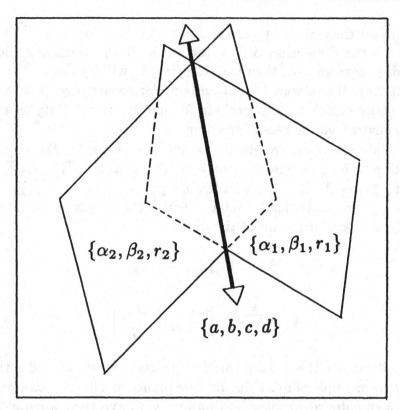

Figure 2.7: A three-dimensional line segment found by project-ing two infinite planes.

Consider an infinite plane corresponding to an observed line segment in an image described by

$$g(\mathbf{x}, \mathbf{p}_i) = x \sin \alpha_i \cos \beta_i + y \sin \alpha_i \sin \beta_i + z \cos \beta_i - r_i = 0$$

parameterized by $\mathbf{p}_i = [\alpha_i, \beta_i, r_i]^T$ with $\mathbf{x} = [x, y, z]^T$ defined in some global coordinate system. A three-dimensional line segment is described by;

$$g(\mathbf{x}, \mathbf{q}) = \begin{bmatrix} x + az - c \\ y + bz - d \end{bmatrix} = 0$$

with $\mathbf{q} = [a, b, c, d]^T$. The equation for this edge can be cal-culated from two observed plane equations $\mathbf{p}_1$ and $\mathbf{p}_2$ by elimination, to obtain the parameter transform $\mathbf{q} = {}^q\mathbf{h}_p(\mathbf{p}_1, \mathbf{p}_2)$. If each plane parameter vector is independent and Gaussian $\mathbf{p}_i \sim N(\hat{\mathbf{p}}_i, \mathbf{\Lambda}_i)$, then the mean value of the resulting edge can be found from

$$\hat{\mathbf{q}} = {}^q\mathbf{h}_p(\hat{\mathbf{p}}_1, \hat{\mathbf{p}}_2)$$

and it's variance from

$$\mathbf{\Lambda}_q = \left(\frac{\partial^q \mathbf{h}_p}{\partial \mathbf{p}_1}\right)\mathbf{\Lambda}_1\left(\frac{\partial^q \mathbf{h}_p}{\partial \mathbf{p}_1}\right)^T + \left(\frac{\partial^q \mathbf{h}_p}{\partial \mathbf{p}_2}\right)\mathbf{\Lambda}_2\left(\frac{\partial^q \mathbf{h}_p}{\partial \mathbf{p}_2}\right)^T$$

This provides an estimate of edge location from two line segments in image space. The dimension of each Jacobian in this case is 4×3, so each single line segment contribution is underconstrained.

The important steps in this transformation of geometric information are to describe the change in parameter vector by the function $\mathbf{h}$, calculate the Jacobian $\mathbf{J}$ and ensure it is well behaved, and to account for the resulting constraints in the new representation. This technique applies to all geometric information described by Equation 2.1.

2.7   Gaussian Topology

The representation and transformation of geometric information must also be coupled with a mechanism to maintain constraints between Gaussian feature descriptions. In this section, we will develop stochastic topology in a Gaussian framework, providing a means to propagate changes in description and maintain consistency between Gaussian features.

In section 2.4.3 we described how constraints between geometric features could be represented by a directed network with arcs labeled by random variables. The three results presented, applied to all possible probability distributions on arc labels e_i. In the Gaussian approximation, we will only be interested in constraints on the mean and variance of these distributions.

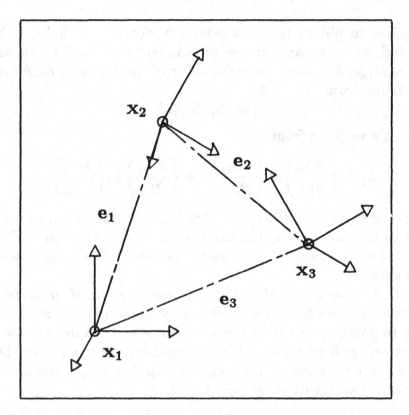

Figure 2.8: A network of three geometric features.

Clearly, the relations on mean values will be the same as those previously developed, therefore our interest in this section is to derive similar constraints on the arc-label variance under the Gaussian assumption.

Consider first the three feature network in Figure 2.8, described by the vector of relations $\mathbf{e} = [\mathbf{e}_1^T, \mathbf{e}_2^T, \mathbf{e}_3^T]^T$. To apply the results of Section 2.4.3 we must first describe each relation in a *common*, but otherwise *arbitrary*, coordinate frame; c^0 for example. This transforms each relation as ${}^0\mathbf{e}_i = {}^0\mathbf{h}_i(\mathbf{e}_i)$. Using the path matrix **C**, Result 2.1 shows that we must have;

$$C[^0h_1(e_1), {}^0h_2(e_2), {}^0h_3(e_3)] = {}^0h_1(e_1) + {}^0h_2(e_2) + {}^0h_3(e_3)$$
$$= {}^0e_1 + {}^0e_2 + {}^0e_3$$
$$= 0$$

with probability 1. Taking expectations, we obtain:

$$^0\hat{e}_1 + {}^0\hat{e}_2 + {}^0\hat{e}_3 = 0 \tag{2.43}$$

If $^0J_i = \frac{\partial {}^0h_i}{\partial e_i}|_{e_i = \hat{e}}$, then the variance matrix on each arc label will be transformed to the new coordinate system c^0 by:

$$^0\Lambda_i = {}^0J_i \Lambda_i {}^0J_i^T \tag{2.44}$$

In this common coordinate system, Result 2.2 requires that:

$$C^0\Lambda C^T = C \begin{bmatrix} ^0\Lambda_1 & ^0\Lambda_{12} & ^0\Lambda_{13} \\ ^0\Lambda_{12}^T & ^0\Lambda_2 & ^0\Lambda_{23} \\ ^0\Lambda_{13}^T & ^0\Lambda_{23}^T & ^0\Lambda_3 \end{bmatrix} C^T = 0 \tag{2.45}$$

Without loss of generality we can assume the cross correlations are symmetric whence:

$$^0\Lambda_1 + {}^0\Lambda_2 + {}^0\Lambda_3 + 2^0\Lambda_{12} - 2^0\Lambda_{13} - 2^0\Lambda_{23} = 0 \tag{2.46}$$

Result 2.3 states that for an arbitrary set of vectors $x = [x_1, x_2, x_3]^T$ we must have $^0e = M^Tx$. As x can be arbitrary, we will choose it so that its joint variance matrix Γ is block diagonal,

$$\Gamma = \begin{bmatrix} \Gamma_1 & 0 & 0 \\ 0 & \Gamma_2 & 0 \\ 0 & 0 & \Gamma_3 \end{bmatrix}$$

We then have:

$$^0\Lambda = M^T\Gamma M = \begin{bmatrix} \Gamma_1 + \Gamma_3 & -\Gamma_1 & -\Gamma_3 \\ -\Gamma_1 & \Gamma_1 + \Gamma_2 & -\Gamma_2 \\ -\Gamma_3 & -\Gamma_2 & \Gamma_2 + \Gamma_3 \end{bmatrix}$$

Together with Equation 2.45 this implies that the following relations must be satisfied:

$$\begin{aligned}
{}^0\Lambda_1 &= -{}^0\Lambda_{12} - {}^0\Lambda_1 \\
{}^0\Lambda_2 &= -{}^0\Lambda_{12} - {}^0\Lambda_{23} \\
{}^0\Lambda_3 &= -{}^0\Lambda_{13} - {}^0\Lambda_{23}
\end{aligned} \tag{2.47}$$

If we now wish to infer ${}^0e_3$ from ${}^0e_2$ and ${}^0e_1$, or equivalently to propagate information through ${}^0e_1$ and ${}^0e_2$, then we assume independence of these vectors (${}^0\Lambda_{12} = 0$) and obtain from Equations 2.46 and 2.47:

$$ {}^0\Lambda_3 = {}^0\Lambda_1 + {}^0\Lambda_2 $$

Reverting to our initial coordinate system, we have:

$$ {}^0J_3 {}^0\Lambda_3 {}^0J_3^T = {}^0J_1 {}^0\Lambda_1 {}^0J_1^T + {}^0J_2 {}^0\Lambda_2 {}^0J_2^T \tag{2.48}$$

Equation 2.48 describes how geometric uncertainty measures can be propagated or combined between different coordinate systems. For example if c^0 were defined with ${}^0h_1(e_1) = e_1$ and hence ${}^0h_3(e_3) = -e_3$, then Equation 2.48 becomes

$$ \Lambda_3 = \Lambda_1 + J_1 \Lambda_2 J_1^T \tag{2.49}$$

The corresponding relation for information is given by:

$$ \Sigma_3 = J_1^{-T}\Sigma_2 J_1^{-1}\left[J_1^{-T}\Sigma_2 J_1^{-1} + \Sigma_1\right]^{-1}\Sigma_1 \tag{2.50}$$

These three-relation results find an important application in maintaining consistency between features observed in different locations. A good example of this is given by Smith and Cheesman [Smith 87], who have considered the problem of a robot observing a beacon and using the resulting over-constraint to reduce the uncertainty in the robots new location (Figure 2.9). This same propagation and constraint mechanism can also be applied to other types of features, used as beacons or landmarks.

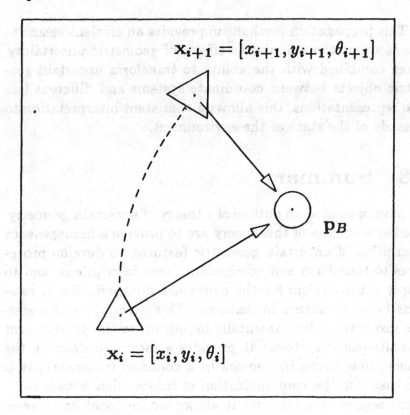

Figure 2.9: A mobile robot observing a beacon from two different locations.

The general Gaussian constraint-network follows the same principles as this three-feature development. However, if there is no change in the vectors e_i during propagation, then all that need be considered is a sequence of these triangular relations (in a similar manner to reducing an electric circuit using Kirchoff's voltage law). If however the vectors do change, then larger networks must be considered. This problem will be encountered when sensor observations must be integrated into an existing prior constraint network, and will consequently be considered along with other changes of constraint in Chapter 4.

This propagation mechanism provides an efficient means to reason about the cumulative effects of geometric uncertainty. When combined with the ability to transform uncertain geometric objects between coordinate systems and different feature representations, this allows a consistent interpretation to be made of the state of the environment.

2.8   Summary

We have presented an outline of a theory of uncertain geometry. The key elements of this theory are to provide a homogeneous description of uncertain geometric features, to develop procedures to transform and manipulate these descriptions, and to supply a mechanism for the consistent interpretation of relations between uncertain features. This description of uncertain geometry is fundamentally important to the development of multi-sensor systems; it provides a means to describe the observations made by sensors in a common framework, it is the basis for the communication of information between different sensor systems, and it allows environment hypotheses suggested by observations to be compared and combined in an efficient and consistent manner. This development of uncertain geometry will be used extensively in later chapters.

Chapter 3

SENSORS AND
SENSOR MODELS

3.1   Introduction

A multi-sensor robot system comprises many diverse sources
of information. The sensors of these systems take observations of a variety of disparate geometric features in the robot
environment. The measurements supplied by the sensors are
uncertain, partial, occasionally spurious or incorrect and often geographically or geometrically incomparable with other
sensor views. This sensor information is inherently dynamic;
the observations supplied by one sensory cue may depend on
measurements made by other sensors, the quality of the information provided may depend on the current state of the robot
environment or the location and state of the sensor itself. It is
the goal of the robot system to coordinate these sensors, direct
them to view areas of interest and to integrate the resulting
observations into a consistent consensus view of the environment which can be used to plan and guide the execution of
tasks.

The sensors used by robot systems are characterized by
the variety of information that they can provide and by the

complexity of their operation. These sensors include vision cameras, tactile arrays, sonar elements, infra-red detectors, laser rangers, thermal imagers, proximity switches, etc. Similar physical devices can often have quite different observation characteristics, depending on their intended use and the environment in which they operate. The information provided by each sensor can be used in many different ways. For example, a visual sensor (camera) acquires a set of raw digitized image scenes of the environment, to this raw information, a diverse range of algorithms can be applied to extract a variety of features; edges, normals, surfaces, centroids, curvature, etc. Indeed, the same information can often be obtained from a given sensor in many different ways; consider for example, the number of edge detector algorithms available for vision systems. Each different source of information, whether a distinct device or algorithm to extract a specific type of feature, can be considered as an individual sensory cue. These cues are observers of the environment, the information they provide is uncertain, it depends on the observations made by other sensors, and is affected by both the state of the environment and the state of the sensor itself. These relations and dependencies may be very complex, but understanding them is essential if we are to utilize the potential abilities of a multi-sensor robot system.

If we are to understand how to fully utilize a sensors capability, we must develop a model of sensor behavior. The purpose of a sensor model is to represent the ability of a sensor to extract descriptions of the environment in terms of a prior world model. Sensor models should provide a *quantative* ability to analyze sensor performance, and allow the development of robust decision procedures for integrating sensor information. The advantage of having a model of sensor performance is that capabilities can be estimated a priori, spurious information can be readily identified, and sensor strategies developed in line with information requirements. In a multi-sensor system, these models can also be used to enhance cooperation between disparate sensory cues and to encourage distribution of sensing and problem solving tasks. Although sensor fusion in a simple homogeneous form can be accomplished without reference to some prior model, such integration methods are difficult to extend and apply to more complex systems, and inherently preclude any reasoning about sensor capabilities.

A sensor model should describes the nature and form of the observations that can be obtained from a particular sensory cue, how this depends on the state and location of the physical device, and how information from other sources affects decisions made with respect to the observations. We can divide a sensor model into three parts; an *observation* model, a *dependency* model, and a *state* model.

An observation model is essentially a static description of sensor performance, it describes the dependence of observations on the state of the environment. For example, if a camera is taking pictures or images of the environment, a static model would describe the ability of the camera to extract edges or surfaces from the image, in terms of locations or feature parameters, together with some measure of their uncertainty. This is the "traditional" observation model used in many other sensor research areas. The construction of observation models is a well developed but still open research issue. In the case of a robot sensor system, observation models are made more difficult because of the often complex way in which the primitive data is built up in to measurements of geometric features.

A dependency model describes the relation between the observations or actions of different sensors. For example, the observations made by a tactile probe may depend on the prior observations supplied by a vision system, or an edge detector may provide a segmentation algorithm with a first estimate of region boundaries. There has been almost no work done on this type of model in either robotics or any other sensor research areas.

A state model describes the dependence of a sensors observations on the location or physical state of a sensing device. For example, a mobile camera platform, may be able to change it's location (viewpoint) or the focus of it's lens to provide different observation characteristics. A description of the dependency of observations on sensor state would enable the development of sensor strategies.

Sensor information is inherently dynamic; observations provided by one sensory cue cannot be considered in isolation from the observations and actions of other sensors. Different sensors may provide quite disparate capabilities, which when considered together are *complementary*. If we have a number of sensors that provide essentially the same information for the purpose of reducing uncertainty and allowing redundant operation, we are likely to have information that is *competitive*, or in disagreement. When two or more sensory cues depend on each other for guidance (multiple-resolution edge detection, or a pre-attentive visual process, for example), then these information sources must *cooperate* to provide observations.

If we are to provide a means to analyze of information in a multi-sensor system, we must develop an effective model of a sensors capabilities, with which to provide an analytic basis for making quantative decisions about sensor organizations. We will develop a probabilistic model of sensor capability in terms of an *information structure*. We consider a multi-sensor system as a team, each observing the environment and making local decisions, each contributing to the consensus view of the world, and cooperating to achieve some common goal. The information structure of a team describes the observations and decisions made by the team members, how they are related, and how they depend on the organization and operation of the team. The concept of a team of sensors provides a powerful means to describe and analyze multi-sensor systems.

In Section 3.2 we will introduce some important characteristics of sensor operation which we will require a sensor model to capture. Section 3.3 describes a number of different methods for modeling multi-sensor systems, and considers how well they describe the characteristics of sensor operation. Section 3.4 introduces the idea of a multi-sensor team. The elements of a team structure are described in the context of distributed sensing and decision making. Starting with a formulation of the team information structure, Section 3.5 develops a team-theoretic model of sensor behavior in terms of a team information structure. This model can be divided in to three parts; observation, dependence, and state models. Section 3.6 considers the observation model, in particular the use of approximate models, with allowance for spurious information. Section 3.7 develops a model of dependency between different sensor observations and actions, in terms of a directed opinion network. Section 3.8 describes a simple model that accounts for the state and location dependence of sensor observations.

All of these models can be placed in a common framework and described by an information structure. These models will be used extensively in the development of strategies for the coordination and control of multiple sensor systems.

3.2   Characterizing Sensors

If we are to provide an accurate model of sensor operation, it is important that we understand and characterize the behavior of the sensing device. We need to describe the way in which information is actually obtained from a sensory cue, how this information is affected by the state of the sensor device, and how the resulting observations can be used by other information sources.

In characterizing a sensors ability to take observations of the environment, we must account for a number of important behaviors:

Device Complexity

When a sensor is composed of many physical devices, each making contributions to the observation extracted by the sensor, the exact distribution of uncertainty and character of the sensor is complex and difficult to describe as an exact model. For example, a camera system comprises a CCD array, lenses, digitizers, etc; their cumulative effect on the sensor character is difficult to predict. This problem is encountered in other sensor-based research areas as well as robotics.

Observation Reconstruction

The feature observations derived from a sensor are in general some complex function of the actual physically measured variables. The reconstruction of geometric observations from such values is often an ill-defined process and has recently received considerable attention in the study of the inverse problem [Poggio 84, Marroquin 85]. A good example of this is the reconstruction of edges from the grey levels of a CCD image. A failure to understand this reconstruction and its effect on sensor characteristics may lead to catastrophic results.

Observation Error

The feature observations reconstructed from sensor data have more uncertainty associated with them than just sensor noise [Durrant-Whyte 87c, Bajcsy 86b]. Error may occur due to inaccuracies in the placement of devices, incorrect interpretation of measurements, or device failure, for example. Uncertainty also arises when a feature observation is incomplete or partial; when information is unavailable in one or more degrees of freedom; all 2D visual cues, for example.

Disambiguation of Features

Many sensors in robotic applications, can take measurements of many features in a single observation. The ability to resolve different features in a single image and to disambiguate them from alternative hypotheses is an important part of a sensors capabilities. In robotics this can be particularly important [Grimson 84], and can lead to a prime source of spurious information. This process is difficult to describe by a conventional noise model and indeed is difficult to model at all. The process of selecting hypothesis to resolve can be computationally expensive and may be further compounded by the complexity of observed features.

Observation Disparity

Robot sensors are characterized by the diversity of observations that can be obtained; edges, normals, locations, or texture for example. If sensor information form many disparate sources is to be combined, we must have an ability to transform one type of uncertain geometric feature to another (edges to surface normals for example). This ability allows observations from different cues to be compared or used to complement each other in deriving a consensus description of the environment.

Multiple Viewpoints

When we have two or more sensors which are geographically separated, we must be able to transform their observations in to a common coordinate system so that they can be compared together. In vision this is called the view-point problem; how to combine observations taken from a number of different viewpoints into a consensus description. The central issue here is to describe the ability of the sensor model to transform and manipulate uncertain descriptions of the environment. This is an important consideration in active sensor descriptions.

Sensor Control

Often robot sensors can be endowed with an ability to actively change their characteristics either by relocating themselves or by changing their internal operation or resolution. This ability, if used properly, can provide an enormously powerful mechanism to refine and resolve the basic observation process, dynamically, in line with system requirements. If such capabilities could be modeled, we could provide a mechanism with which to decide on sensor strategies for obtaining the most relevant information from the environment given the limitations of the sensor mechanism.

Sensor Coordination

One of the advantages of using multiple sensors is to provide overlapping or redundant information sources. To make full use of such facilities we must understand how the decisions made by one sensor depend on the outcome of the decisions made by other sensors. This interaction between information sources is intrinsically dynamic. These interactions require us to understand how disparate information sources might cooperate or compliment each other, and why they may disagree with each others decisions.

To provide a homogeneous method for describing these diverse characteristics, we will develop a model of sensor observations as a probability distribution on the parameter vector of a measured feature. This enables us to account for noisy, partial information, and to develop integration procedures robust to spurious measurements, modeling inaccuracies and mistakes. In conjunction with the description of uncertain geometry presented in Chapter 2, dynamic dependencies between sensors can be provided for in a simple and conceptually transparent manner. Probabilistic models of sensor capabilities have a number of advantages:

- Probabilistic models are well suited to describing the inherent uncertainty characteristic of making observations in the real world.

- Well-tested methodologies exist for analyzing observations described as probability distributions.

- The description of different cues in a common probabilistic framework allows diverse observations to be compared and integrated in a consistent manner.

- If a common modeling policy is used, it becomes very easy to extend the system by adding new sensors.

In the following sections we will develop these probabilistic models in terms of the team information structure.

3.3 Multi-Sensor System Models

The realization that a multi-sensor system must be modeled to be understood is not a new idea. However, this need has often been subsumed in to an implicit recognition of the limitations of a sensor with respect to some prescribed task. This approach to modeling leads to impossibly complex integration procedures and an inability to understand or utilize the power available from a multi-sensor system.

The most comprehensive rationalization of sensor abilities in the context of a multi-sensor system is probably Henderson's logical sensor descriptions [Henderson 84,85]. These logical sensor models[1] provide a description of sensor capabilities in terms of a set of *rules* specifying input-output characteristics of a device or algorithm. These models can be built up in different ways to obtain a description of how information from sensors can be used to satisfy system demand. This logical sensor system basically provides a programming environment in which sensor algorithm capabilities can be described and manipulated.

Another rule-based sensor model has been proposed by Flynn [Flynn 85,87] for integrating sonar and infra-red sensor information as applied to mobile robot map-building. Sonar and infra-red sensors provide a very good compliment to each others abilities; infra-red has good angular resolution properties, but poor depth-measuring ability, whereas sonar provides

---

[1] Henderson denies that these descriptions are "models" of sensor ability.

comparatively good depth observations but with poor angular resolution. Together, they can supply a comprehensive (and cheap) navigation map. Flynn has devised four rules which describe when either sonar or infra-red sensor information is to be used to build the map. The complimentary nature of these two information sources (the two sets of information were never integrated at a single location) allowed this model of capability to function efficiently.

There are a number of overriding problems with these rule-based approachs to sensor modeling: They provide no explicit means of representing or reasoning with uncertainty, they preclude any quantative analysis of system performance, and they are difficult to extend in any consistent, homogeneous fashion. We want to take a far more general approach to sensor modeling than that provided by rule-based systems. We want to be able to consider arbitrary sequences of observations made by diverse sensor systems, be able to integrate these observations, use them to coordinate and control sensing devices, and to provide a quantative means of estimating system performance.

To fully utilize the capabilities of many robot sensors we must also be able to characterize the dynamic nature of observations; interdependencies between information sources and the effect of sensor state on the observations supplied. We can delineate three basic types of information interaction:

1. *Complimentary* information, when two disparate sources provide information which together provide a more global view of the environment but which do not overlap by themselves.

2. *Competitive* information, when two sources provide the same piece of information and a joint consensus must be reached.

3. *Cooperative* information, when one information source

must rely on or cooperative with another in order to obtain observations.

Examples of the dynamic nature of observations occurs at all levels of sensor modeling: For example, at the lowest level between coarse and fine disparity arrays in Marr's model of early vision, at an intermediate level between edge and surface models, and at the highest level between sensor motions. It is not at all clear how these models should be developed. There have been some solutions proposed for specific dynamic problems. Trezopolous [Trezopolous 86] has demonstrated a variational approach for integrating complimentary edge and surface normal constraints to improve a finite element, visual-surface reconstruction algorithm. Allen [Allen 85] has demonstrated the use of a tactile sensor to supplement the information provided by a stereo camera system; the tactile sensor investigating occluded parts in a complimentary rather than cooperative manner. Neither of these two techniques approach the general problem of sensor interaction.

3.4 Sensors as Team Members

We will consider a multi-sensor system as a team of decision makers. The sensors are members of the team, each observing and contributing to the team-consensus view of the environment. Each individual sensor can make local decisions based on its own observations, but must cooperate with the other team members to achieve a common objective. A team structure has a number of important characteristics: Team members help each other by exchanging information, thus improving the team as a whole. Team members can disagree with each other as to what is being observed, thus supplying a mechanism for validating each others operation and resolving differences of opinion. Team members can supply each other with information unattainable to individual members, thus providing the team with a more complete description of events. The most important observation about a team is that the members acting *together* are more effective than the sum of their individual actions.

The characteristics of a team capture many of the desirable properties of a multi-sensor system. An important element of this similarity is the ability of individual sensors to communicate information to other team members. To communicate, sensors need to understand a common language. We will choose the stochastic geometry described in Chapter 2 as this common language. The exchange of information provides a basis through which individual sensors can *cooperate* with each other, resolve *conflicts* or disagreements, or *compliment* each others view of the environment.

Team decision theory was originally developed to provide a quantative mechanism for describing economic organizations. In their book, Marshak and Radnor [Marshak 72], describe the extension of single person decision problems to multi-person organizations. The primary emphasis of this work is to provide a means of coordinating a distributed decision making process. Each decision maker is only allowed access to partial state information (as a model of bounded rationality [Simon 62]), from which local decisions must be made, reflecting some global optimality criteria. This team structure has also been developed as a means of describing problems in multi-person control. In a series of papers, Ho and Chu [Ho 72,79,80, Chu 72] have investigated the distributed linear quadratic Gaussian control problem, as applied to a group of sequential decision makers.

We propose to describe a multi-sensor system as a team in the following sense: The sensors are considered as members of the team, each observing the environment and making local decisions based on the information available to them. The observations z made by a sensor are described by an information structure η. Each sensor can make a decision δ, based on these observations, resulting in an action a, usually an estimate of a feature description, describing the opinion of the sensor. The opinions of each sensor are integrated to provide a team decision and action.

The ith team member is described by an *information struc-ture* η_i. This structure is a model of sensor capabilities. An information structure is defined by the relation between obser-vations, state and decisions;

Definition: The *information structure* of the ith sensor or team member $(i = 1, \cdots, n)$ is a function η_i which describes the observations z_i made by a sensor in terms of it's physi-cal state x_i, available prior information about the state of the environment $p_i \in P_i$, and the other sensors or team members actions $a_j \in A_j, j = 1, \cdots, n$. So that:

$$z_i = \eta_i(x_i, p_i, a_1, \cdots, a_{i-1}, a_{i+1}, \cdots, a_n) \qquad (3.1)$$

☐

Collectively the n-tuple $\vec{\eta} = (\eta_1, \cdots, \eta_n)$ is called the infor-mation structure of the team. The action a_i of the ith team member is related to its information z_i by a decision function $\delta \in D_i$ as $a_i = \delta_i(z_i)$. Collectively the n-tuple $\vec{\delta} = (\delta_1, \cdots, \delta_n)$ is called the team decision function. We will return to the decision problem in Chapter 5.

The observations z_i provided by each sensor are essentially estimates of some set of geometric quantities $p_i \in P_i$ (edges, surface normals, etc) in the environment. The information structure is intended to describe the observations obtained from a sensor in terms of these geometric features.

There are a number of different forms that the information structure can take. If for all team members η_i is defined *only* on the prior information p_i then the resulting structure is called a *static* team [Marshak 72] and corresponds to a pure observation model. When η_i also depends on the other teams actions then the structure is called a *dynamic* team [Ho 72]. As each team member can not make decisions and be aware of the result simultaneously, the general form of information structure for a dynamic team must induce a causal relation on the members actions a_j. There are two ways to achieve this: We can apply a precedence structure on the time instant a member makes a decision, so that if member i makes a decision prior to member j then the information structure η_i will not be a function of a_j. Indexing the team members by their decision making precedence order we can rewrite the information structure as:

$$z_i = \eta_i(x, p_i, a_1, \cdots, a_{i-1})$$

This is called a non-recursive team structure (see Figure 3.2 for example). Alternatively, the team members may make decisions asynchronously, conducting a dialogue, such that their actions are indexed by time. Then with $a_i^{t+1} = \delta_i(z_i^{t+1})$ the information structure can be written as:

$$z_i^{t+1} = \eta_i(x, p_i, a_1^t, \cdots, a_n^t)$$

This is called a recursive team structure or Markovian dialogue [Bachrach 75][2] (see Figure 3.2 for example). Under certain conditions the recursive exchange of the teams opinions can be shown to converge to a consensus [Bachrach 79].

   This dialogue may also be written in the form of a precedence relation. If η_i is such that knowing the jth action also implies knowledge of the jth members observation z_j, then the resulting structure is called "partially nested". This structure has the property that knowledge of a team members decisions also implies knowledge of the observation.

   We will be concerned with the model of sensor performance provided by the information structure. We consider three types of sensor model, an *observation* model, a *dependence* model and

---

[2]It is always possible to reduce a recursive team structure to a non-recursive form by considering each timed action as the decision of a new decision maker.

state model. The observation model η_i^p describes the character of measurements given the state of the sensor and all other sensor actions. The dependence model η_i^δ describes the effect of other sensor actions (observations) on the sensor measurements. The state model η_i^x describes the observation dependence on the internal state and location of the sensor. We will embed all these models in a common information structure η_i.

We are primarily interested in teams of observers, where the action a_i of each sensor is to make an estimate of some geometric feature $a_i \in P$ in the environment. By relating the actions of each sensor or team member to it's observations through a decision function $\delta_i(z_i) \mapsto a_i$, we can then rewrite the general information structure as:

$$z_i = \eta_i(x_i, p_i, \delta_1(z_1), \cdots, \delta_{i-1}(z_{i-1}), \delta_{i+1}(z_{i+1}), \cdots, \delta_n(z_n))$$
$$(3.2)$$

This allows us to consider the information structure as a transformation of observations to observations; the observations z_i are random vectors and η_i is a stochastic function transforming the decisions $\delta_j(\cdot) \in P_j$, the prior information $p_i \in P_i$ and the sensor state x_i into elements of P.

This transformation of observations between different sensors or team members suggests that we describe each decision function in terms of a distribution function, so that

$$f(z_i)$$

$$= f(\eta_i(x_i, p, \delta_1(z_1), \cdots, \delta_{i-1}(z_{i-1}), \delta_{i+1}(z_{i+1}), \cdots, \delta_n(z_n)))$$

$$= f_{\eta_i}(x_i, p, \delta_1(z_1), \cdots, \delta_{i-1}(z_{i-1}), \delta_{i+1}(z_{i+1}), \cdots, \delta_n(z_n))$$

This representation of the information structure has a number of advantages: The description of observations is now dimensionless so we can manipulate this function using standard probabilistic techniques and apply our development of uncertain geometry to the communication of information between different sensory cues. The decisions $\delta_i(\cdot)$ can be related through f_{η_i} to any common decision philosophies; maximum likelyhood, etc. This allows us to develop decision procedures capable of using many disparate sources of information (see Chapter 5). Another important advantage of describing the information structure in terms of a probability distribution function is that the variables (prior information, state and other sensor decisions) can be separated from each other by expanding f_{η_i} as a series of conditional probability distributions. This in turn allows us to decouple the three types of sensor model, $(\eta^p, \eta^\delta, \text{and } \eta^x)$ from each other. Let

$$\overline{\delta}_i = (\delta_1, \cdots, \delta_{i-1}, \delta_{i+1}, \cdots, \delta_n)$$

then;

$$
\begin{aligned}
f(\mathbf{z}_i) &= f_{\eta_i}(\mathbf{x}_i, \mathbf{p}_i, \overline{\delta}_i) \\
&= f_{\eta_i}(\mathbf{x}_i \mid \mathbf{p}_i, \overline{\delta}_i) f_{\eta_i}(\mathbf{p}_i \mid \overline{\delta}_i) f_{\eta_i}(\overline{\delta}_i) \qquad (3.3) \\
&= f_x(\eta_i^x) f_p(\eta_i^p) f_\delta(\eta_i^\delta)
\end{aligned}
$$

The state model $f_x(\eta_i^x)$ now describes the dependence of a sensors observations on its location and internal state *given* any prior information and all other sensor opinions. The observation model $f_p(\eta_i^p)$ describes the dependence of sensor measurements on the state of the environment *given* all other sensor decisions. The dependence model $f_\delta(\eta_i^\delta)$ describes the prior information supplied by the other sensors in the system. The product of these three models describes the observations made by the sensor.

This ability to decouple sensor models enables the effect of different types of information to be analyzed independently. This in turn provides a powerful framework in which to develop descriptions of sensor performance.

3.5   Observation Models

All geometric features in the environment are modeled by functions $\mathbf{g}(\mathbf{x}, \mathbf{p}) = 0$. Each function defines a family of features, parameterized by the vector **p**. Our world model characterizes each feature by a probability density function $f_g(\mathbf{p})$ on this parameter vector. Consider a sensor taking observations of a geometric feature, an instance of a given family parameterized by the vector **p**. We can model this observation as a conditional probability distribution $f_g(\mathbf{z} \mid \mathbf{p})$ that describes the likelyhood of feature observation given all prior information about **p**. This distribution is our observation model; $f_g(\mathbf{z} \mid \mathbf{p}) \equiv f_p(\eta^p)$.

The exact form of $f(\cdot \mid \mathbf{p})$ will depend on many physical factors. It is unlikely that we can obtain an exact description of the probabilistic character of observations in all but the simplest of cases. It may in fact be undesirable to use an exact model even if it were available because of its likely computational complexity, and its inability to model non-noise errors such as software failures or algorithmic misclassifications. It is usual to assume a Gaussian or Uniform distribution model for the conditional observations [Faugeras 85]. These models allow the development of Computationally simple decision procedures. However they are not able to represent poor information and can fail with catastrophic results even when the observations deviate only a small amount from the assumed model. In the absence of an exact observation description, we propose to use an approximate model, coupled with a decision procedure robust to model inaccuracies.

There are two different ways to approach this approximation problem; we can model the observation noise as a *class* of possible distributions, or as some nominal distribution *together* with an additional unknown likelyhood of errors and mistakes. McKendall and Mintz [McKendall 87] have conducted experiments on a stereo camera system, and as a result have proposed modeling visual noise as a class of probability distributions. Applying a minimax[3] philosophy to this class results in a soft-

---

[3]Using a maximum à posteriori estimator with the most pessimistic distribution form a class of distributions.

quantizer as a decision procedure [Zeytinoglu 84]. In the long term these models show great promise, however they currently have two major problems; the relative computational complexity of the decision procedures, and their inability to describe non-noise errors.

We require a computationally tractable, approximate sensor model which performs efficiently when close to the assumed model, which is robust to small deviations in misspecification of the model, and which will not result in catastrophe when large deviations are experienced. We propose to use an approximation that describes the observations by some nominal distribution together with an unknown likelyhood of errors or mistakes. Then we shall require that our decision procedures be robust to the possibility of error, but otherwise assume a nominal noise model. These distributions are termed *gross error models* [Huber 81], and have the general form:

$$P_\epsilon(F_0) = \{F \mid F = (1 - \epsilon)F_0 + \epsilon H, \quad H \in M\} \qquad (3.4)$$

These models are described by a set of distributions F which are composed of some nominal distribution F_0 together with a small fraction ϵ of a second probability measure H. This second measure is often assumed unknown and acts to contaminate the nominal distribution with unexpected observations. Statistically the purpose of the contamination is to flatten the tails of the observation distributions and force resulting decision procedures to be robust with respect to possible outlying measurements. Ignoring the flattening of the distribution tails can cause catastrophic results in apparently quite reasonable cases (for example, if a Gaussian model is assumed when the observations are in reality derived from a Cauchy distribution). This model results in decision procedures which cluster observations and trim outliers from consideration in the integration process. If the nominal model has finite moments, then this clustering process converges to a Gaussian observation model. For this reason, we consider a particular case of the gross error

model called the contaminated Gaussian distribution, which
has the general form;

$$f(\mathbf{z} \mid \mathbf{p}) = \frac{(1-\epsilon)}{(2\pi)^{\frac{m}{2}}|\mathbf{\Lambda_1}|^{\frac{1}{2}}} \exp\left[-\frac{1}{2}(\mathbf{z} - \mathbf{p})^T\mathbf{\Lambda_1}^{-1}(\mathbf{z} - \mathbf{p})\right]$$
$$+ \frac{\epsilon}{(2\pi)^{\frac{m}{2}}|\mathbf{\Lambda_2}|^{\frac{1}{2}}} \exp\left[-\frac{1}{2}(\mathbf{z} - \mathbf{p})^T\mathbf{\Lambda_2}^{-1}(\mathbf{z} - \mathbf{p})\right] \quad (3.5)$$

with $0.01 < \epsilon < 0.05$ and $|\mathbf{\Lambda_1}| \ll |\mathbf{\Lambda_2}|$. The spirit of this model
is that the sensor behaves as $N(\mathbf{p}, \mathbf{\Lambda_1})$ most of the time but sub-
mits occasional spurious measurements from $N(\mathbf{p}, \mathbf{\Lambda_2})$. With
the type of sensors that we are considering (vision, tactile,
range, etc), the contaminated model is intended to represent
the fact that we would normally expect quite accurate observations within a specific range, but must be robust to problems
like miscalibration, spurious matching and software failures.
We assume that we have knowledge of the value of $\mathbf{\Lambda_1}$ from the
character of the sensor, but we do not explicitly assume any-
thing other than bounds on the values of $\mathbf{\Lambda_2}$ and ϵ. This has
the property of forcing any reasonable integration policy devel-
oped to be robust to a wide variety of sensor observations and
malfunctions. This model is thought to be a sufficiently con-
servative estimate of sensor behavior, in that by choosing $\mathbf{\Lambda_1}$,
$\mathbf{\Lambda_2}$ and ϵ sufficiently large, we can encompass all the possible
uncertainty characteristics we may expect to encounter. The
intention of this model is to approximate a sensors true characteristics without having to analyze possible sensor responses.
This forces us to develop decision procedures which are robust
to the exact specification of observation model, which provide
efficient results in the light of our poor knowledge of sensor
characteristics, and that will be robust to spurious or gross
contaminations.

The conditional density function $f(\mathbf{z} \mid \mathbf{p})$ represents our
observation model of the sensor:

$$f(\mathbf{z} \mid \mathbf{p}) = f_{\eta_i}(\mathbf{p} \mid \overline{\delta}_i) \quad (3.6)$$

This describes the distribution of observations given all other prior information about sensor location and other sensor states. The observation model developed here describes the distribution of a single feature measurement. We must also be concerned with multiple-sample characteristics, the statistical correlation between observations, the likely density of observations and the effect of algorithms that aggregate observations in to higher level environment descriptions. We maintain that these issues should be considered in terms of decision procedures rather than in the sensor model per se. In the following we will use the ϵ-contamination model to develop decision rules which are sympathetic to the *actual* character of sensor observations and which are robust with respect to spurious information and model specification.

3.6 Dependency Models

If we are to fully understand and utilize sensor capabilities, it is important that we model the dynamic nature of sensor information. Unlike the development of observation models, there has essentially been no research in dynamic sensor models in robotics or any other sensor based research areas. We will develop a model of dependence between sensory systems by considering the dependence model η_i^δ in terms of a set of conditional probabilities on the information provided by other sensors and cues.

Consider the dependence of the ith sensor or team members information structure as described by the dependence model:

$$\eta_i^\delta(\overline{\delta}_i) = \eta_i^\delta(\delta_1(z_1), \cdots, \delta_{i-1}(z_{i-1}), \delta_{i+1}(z_{i+1}), \cdots, \delta_n(z_n)) \quad (3.7)$$

We will interpret this dependency model as a distribution function, describing the information provided to the ith sensor by all other sensors as a prior probability of feature observation:

$$f_i(\eta_i^\delta(\overline{\delta}_i)) = f_\delta(\overline{\delta}_i(z_1, \cdots, z_{i-1}, z_{i+1}, \cdots, z_n))$$

$$= f_\delta(\overline{\delta}_i)$$

(3.8)

This interpretation makes statistical sense as the joint feature density is found by multiplying the conditional observation model $f_i(z \mid p) = f_p(p \mid \overline{\delta}_i)$ by the prior information $f_i(p) = f_\delta(\overline{\delta}_i)$, so that following Bayes-rule:

$$f_i(z, p) = f_i(z \mid p)f_i(p)$$

$$= f_p(p \mid \overline{\delta}_i)f_\delta(\overline{\delta}_i)$$

(3.9)

$$= f_\eta(\eta_i(p, \overline{\delta}_i))$$

This interpretation is also intuitively appealing as it is the observations made by other sensors, communicated to the ith sensor, that provide the initial prior information. It is worth noting that in this sense, there is no difference between prior information supplied by a sensor and that supplied by an existing world model.

The internal structure of the distribution $f_\delta(\cdot)$ must, in general, describe the interdependence of all sensor information contributing to the ith information structure. These dependencies should represent the sequence in which information is passed between sensors, and describe the way in which observations combine in complimentary, cooperative and competitive ways to provide a full and complete environment description. Each observation z_i made by a sensor is a measurement of a specific type of feature $p_i \in P_i$, the decision $\delta_i \in P_i$ is an estimate of this feature based on the observation; the parameter vector of an edge or surface equation, for example. The information structure η_i is an interpretation of this estimate in terms of a distribution function defined on the parameter vector. The dependency model $\eta_i^\delta(\cdots, \delta_j, \cdots)$ is therefore a *trans-formation* of feature descriptions obtained by the jth sensor into feature descriptions required by the ith sensor. This transformation represents a change in stochastic feature descriptions. It is best described in terms of probability distribution functions, allowing use of the techniques described in Chapter 2 to effect the transformation of disparate sensor opinions.

The interpretation of the dependency model as a probability distribution allows us to expand $f_\delta(\cdot)$ as a series of conditional distributions describing the effect of each individual sensor on the ith sensors prior information. For example, if the numeric order of decision making is also the natural precedence then;

$$f_\delta(\overline{\delta_i}) = f_\delta^i(\delta_1, \cdots, \delta_n)$$

$$= f_\delta^i(\delta_1 \mid \delta_2, \cdots, \delta_n) f_\delta^i(\delta_2 \mid \delta_3, \cdots, \delta_n) \cdots f_\delta^i(\delta_n)$$

$$= f(\eta_1^i(\delta_1) \mid \eta_1^\delta(\delta_2, \cdots, \delta_n)) \cdots f(\eta_n^i(\delta_n))$$

Each term $f_\delta^i(\delta_j \mid \delta_k) = f(\eta_j^i(\delta_j) \mid \eta_j^\delta(\delta_k))$ describes the information contributed by the jth sensor to the ith sensors prior information, given that the information provided by the kth sensor is already known. The transformation affected by $\eta_j^i(\delta_j)$ takes the jth sensor observation and interprets it in terms of observations made by the ith sensor; $\eta_j^i(\cdot) \in P_i$. The term $\eta_j^\delta(\delta_k)$ is the dependence model describing the kth sensors contribution to the ith sensors prior information. This in turn can be described by the probability distribution $f_\delta^j(\delta_k)$. If the jth sensor does not depend directly on any information provided by the kth sensor, then f_δ^j can be considered non-informative;

$f_\delta^j(\delta_k) = 1$. Consequently;

$$
\begin{aligned}
f_\delta^i(\delta_j \mid \delta_k) &= f_i(\eta_j^i(\delta_j) \mid \eta_j^\delta(\delta_k)) \\
&= f_i(\eta_j^i(\delta_j)) \qquad\qquad (3.10) \\
&= f_\delta^i(\delta_j)
\end{aligned}
$$

This decomposition by conditionals can, in general, be written in any appropriate order. If there is a natural precedence order in which sensors take observations, then an expansion in that order is appropriate. For example, if the decision δ_j only depends on information provided by the $(j-1)^{th}$ sensor, then the ith dependence model can be represented by;

$$f_\delta^i(\delta_1, \cdots, \delta_{i-1}) = f_\delta^i(\delta_{i-1} \mid \delta_{i-2}) \cdots f_\delta^i(\delta_2 \mid \delta_1) f_\delta^i(\delta_1) \qquad (3.11)$$

describing a Markovian chain of decision makers[4], as shown in Figure 3.1.

The use of conditional probability distributions in this way induces a *network* structure of relations between different sensors and cues. This network is a constraint exposing description of sensor capabilities: Each arc in the network describes a dependence between sensor observations, represented by a constraining transformation and implemented by the propagation of prior information between sensors. Each decision making node or sensor uses this prior information to guide or constrain the acquisition of new observations. The network induced by the expansion of the sensor dependence model has much in common with the constraint network describing relations between geometric features as developed in Chapter 2. This is not unexpected, as each of our sensory systems is considered

---

[4]For example, a pyramid architecture for image processing allows the use of results from coarse image analysis to guide finer resolution processing, a particular level in the architecture getting prior information from the immediate preceding level.

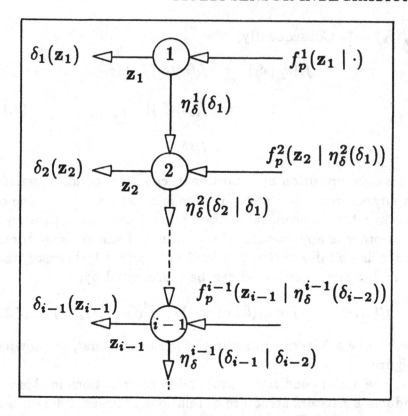

Figure 3.1: The Markovian team decision network

to extract the parameters of a geometric feature, the relations between sensors should just be the relations between features.

The interpretation of the exchange of sensor opinions as a network with arcs labeled by probability distributions is intended to describe how information from one sensor can be used to drive the extraction of information from another sensor. The dynamic use of information may occur in a number of ways, all of which can be classified in terms of either competitive, complimentary or cooperative information intercation. The dependence model can be used to describe all three of these dynamic effects:

- *Competitive Information* interaction occurs whenever two

(or more) dependence models supply information in the same location and degrees of freedom. In this case

$$f_\delta^i(\delta_1, \delta_2) = f^i(\delta_1) f_\delta^i(\delta_2)$$

is the joint information. For a consensus decision to be made, f_δ^i must be unimodal (Chapter 5), otherwise differences of opinion must be resolved by resort to further information. Note that δ_1 and δ_2 can describe quite different geometric features, but when transformed by f_δ^i to a common description, can still be competitive. Examples of competitive information interaction include any feature clustering algorithms or any homogeneous observation integration problem (Chapter 4).

- *Complimentary Information* interaction occurs whenever two (or more) information sources supply different information about the same geometric feature, in different degrees of freedom. In this case, the joint information is composed from different non-overlapping information sources. It follows that each dependence model $f_\delta^i(\delta_j)$ must only supply partial information about the ith sensors observations. A simple example of this would be the different beacons of a triangulation system. A more complex example might be the use of a tactile sensor to fill in observations occluded from a vision system [Allen 85]. Complimentary information can often be considered in terms of one sensor "filling in the gaps" in another sensors observations.

- *Cooperative information* interaction occurs whenever one sensor relies on another for information, prior to observation. In this case,

$$f_\delta^i(\delta_1, \delta_2) = f_\delta^i(\delta_1 \mid \delta_2) f_\delta^i(\delta_2)$$

describes the joint information. Pure cooperation, $f_i^\delta(\delta_1 \mid \delta_2) = 1$, is not very interesting. More important is the use of one sensors information to guide the search for new observations. A multi-resolution edge-detector would be a simple example of this process, the guiding of a tactile senor by initial visual inspection would be a more complex example.

There are two distinct processes going on in each of these three types of interaction; the local decision made by each sensor, and the transformation of this information between decision making nodes. Each sensor takes observations described by the observation model $f_i^p(\mathbf{p} \mid \cdot)$, and is provided with prior information from other sensors through the dependence model $f_i^\delta(\cdot)$. The sensor makes a decision $\delta_i(\cdot)$ based on the product of observed and prior information. This is passed on to the next sensor in terms of it's dependence model $f_{i+1}^\delta(\delta_i(\cdot), \cdots)$, which transforms the information provided by δ_i into the feature description understood by δ_{i+1}. This transformation and combination of information from disparate sources demonstrates the clear advantage of describing information in terms of a dimensionless probability distribution function.

There are basically two types of sensor network; the recursive and non-recursive forms. A non-recursive network (Figure 3.2) contains no loops, so that the flow of information between sensors is always in one direction. In a recursive network (Figure 3.3), loops exist, allowing two-way communication between sensors. The advantage of a non-recursive network is that the natural precedence order on sensors makes the analysis of the process of communication transparent. The advantage of recursive networks is that sensors can engage in a dialogue, successively suggesting and verifying each others observation hypotheses, providing, through mutual constraint, a powerful means to resolve an environment description. In a non-recursive network, observations are propagated from one sensor to another through the constraint described by transformation. The constraint is due to the topological character

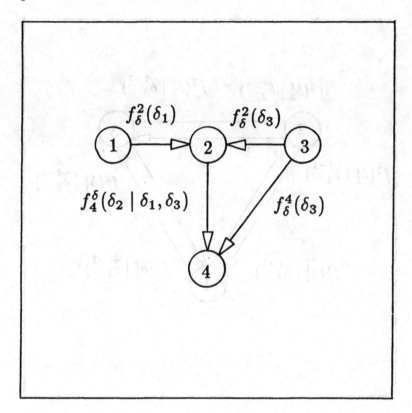

Figure 3.2: A non-recursive network of sensor systems

of the environment geometry, constraining and reducing sensor opinions. The propagation is due to the broadcasting of information from one sensor to others, guiding the search space and resolution process. In a recursive sensor network this propagation and constraint reduces to a dialogue between sensors: A sensor produces uncertain and partial information which is propagated through the network of opinions. Other sensors take this information and use it to guide their own observations which in turn constrain the possible interpretations that can be given to a consensus decision. The results of these constrained observations can then be returned to the original sensor, initiating another round of discussions.

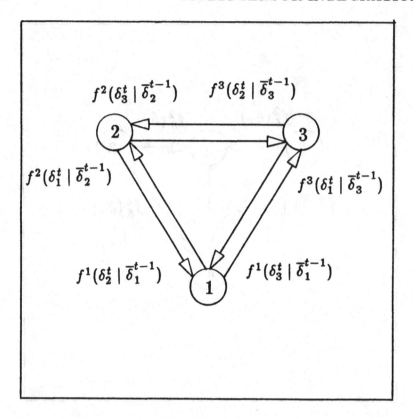

Figure 3.3: A recursive network of sensor systems

To fix these ideas we will consider a simple (non-recursive) example of a vision system (1), seeking to find planer surfaces, using a two-resolution edge detector (2,3) and a region-growing algorithm (4) (Figure 3.4), described by

$$f_1^\delta(\delta_2, \delta_3, \delta_4) = f_1^\delta(\delta_4 \mid \delta_3) f_1^\delta(\delta_3 \mid \delta_2) f_1^\delta(\delta_2) \qquad (3.12)$$

The high resolution edge detector receives information from the lower resolution edge finder described by the information structure $f_1^\delta(\delta_2)$. In this case the lower resolution detector is just used to provide preliminary estimates for use by the fine edge detector; ie prior information. The region-grower uses the information provided by the fine edge detector to localize pos-

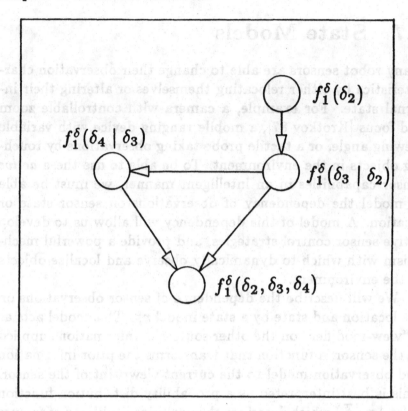

Figure 3.4: A non-recursive network describing a vision system comprising two edge detectors and a region-growing algorithm.

sible surfaces. The information structure $f_4^\delta(\delta_3, \delta_2)$ describes the transformation of edges in to surface representations, and $f_1^\delta(\delta_4 \mid \delta_3)$ describes the observations provided by the region grower. The final result or estimate of surfaces can be found by combining the outputs from both the edge detectors $f_1^\delta(\delta_2, \delta_3)$ and the region-growing algorithm $f_1^\delta(\delta_4 \mid \delta_3)$.

These dependency models and opinion networks provide a powerful means of describing the dynamic exchange of information between different sensor systems. These models will be developed further in the context of the decision problem in Chapter 5.

3.7 State Models

Many robot sensors are able to change their observation characteristics by either relocating themselves or altering their internal state. For example, a camera with controllable zoom and focus [Krotkov 87], a mobile ranging device with variable viewing angle, or a tactile probe taking observations by touching objects in the environment. To be able to use these *active* sensor capabilities in an intelligent manner, we must be able to model the dependency of observations on sensor state or location. A model of this dependency will allow us to develop active sensor control strategies, and provide a powerful mechanism with which to dynamically observe and localize objects in the environment.

We will describe the dependence of sensor observations on it's location and state by a state model η_i^z. This model acts as a "view-modifier" on the other sources of information supplied to the sensor; a function that transforms the prior information and observation model to the current viewpoint of the sensor. This is best interpreted as a probability distribution function $f_z(x_i \mid p_i, \overline{\delta}_i)$ which describes the posterior likelihood of feature observation in terms of a state vector x_i, given the prior information provided by other sensors. This density function can be considered as a transformation that modifies the observed model $f_p(p_i \mid \overline{\delta}_i)$ and the prior information provided by other sensors $f_\delta(\overline{\delta}_i)$ to account for sensor state. The transformation is just a product of distribution functions:

$$f_i(z_i) = f_z(x_i \mid p_i, \overline{\delta}_i) \cdot \left[f_p(p_i \mid \overline{\delta}_i) f_\delta(\overline{\delta}_i) \right] \qquad (3.13)$$

There are two related parts to this description of a state model; the dependency of observation uncertainty on sensor state, and the transformation of prior world information to the current sensor location to determine if a feature is in view. Both of these considerations involve a process of transforming feature descriptions between coordinate systems.

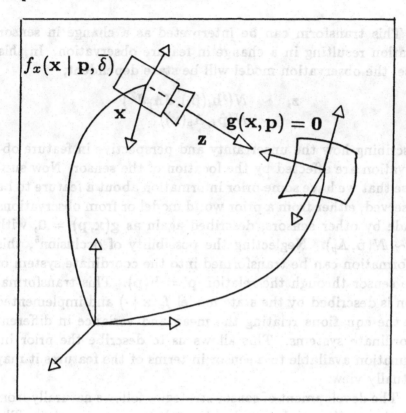

Figure 3.5: A mobile sensor observing a single geometric feature

Consider a mobile sensor located arbitrarily in space by the vector $\mathbf{x} = [x, y, z, \phi, \theta, \psi]^T$, observing a given feature $\mathbf{g}(\mathbf{x}, \mathbf{p}) = 0$ parameterized by the vector $\mathbf{p}$ as shown in Figure 3.5. Suppose that the observations made by this sensor when in a fixed location can be described by a Gaussian distribution $z_i \sim N(\hat{\mathbf{p}}, \Lambda_p)$. This is our observation model. Following Chapter 2, when a Gaussian feature is transformed between coordinate systems as $\mathbf{p}_j = {}^j\mathbf{h}_i(\mathbf{p}_i)$, the mean vector follows the usual laws of geometry, $\hat{\mathbf{p}}_j = {}^j\mathbf{h}_i(\hat{\mathbf{p}}_i)$ and with ${}^j\mathbf{J}_i = \frac{\partial {}^j\mathbf{h}_i}{\partial \mathbf{p}_i}$, the variance matrix is transformed by

$$ {}^j\Lambda_p = {}^j\mathbf{J}_i(\mathbf{x}){}^i\Lambda_p{}^j\mathbf{J}_i^T(\mathbf{x}) $$

This transform can be interpreted as a change in sensor location resulting in a change in feature observation. In this case, the observation model will be state dependent;

$$z_i \quad \sim \quad N(^jh_i(\hat{p}_i), J\Lambda_p J^T)$$
$$\sim \quad N(\hat{p}_j, \Lambda_p(x))$$

describing how the uncertainty and perspective in feature observation are affected by the location of the sensor. Now suppose that we have some prior information about a feature to be observed, either from a prior world model or from observations made by other sensors, described again as $g(x, p) = 0$, with $p \sim N(\hat{p}, \Lambda_p)$. Neglecting the possibility of occlusion[5], this information can be transformed into the coordinate system of the sensor through the relation $p' = h(p)$. This transformation is described by the state model $f_x(x \mid \cdot)$ and implemented by the equations relating the mean and variance in different coordinate systems. This allows us to describe the prior information available to a sensor in terms of the features it may actually view.

The development of sensor strategies follows naturally from this description of observation dependence on sensor state. The senor observation model described in terms of the mean and variance of sensor observations, is now a function of the state and location of the sensor. If, for example, information about some particular feature is required, then this dependence of feature description on state can be used to determine the value of x which would give the measurements made by the sensor some appropriate characteristics. For example, knowing the function $\Lambda_p(x)$, we can proceed to find the senor state x that minimizes (in some sense) the uncertainty in taking observations of p. An example of this type of strategy appears in Chapter 5 in the context of sensor control.

---

[5]The problem of occlusion can be accounted for using standard computer graphics techniques for deciding if a surface is in view from a particular location.

A second important problem in modeling state dependencies is describing the effect on a sensors observations of *other* sensor states. This most obviously occurs when two or more sensors obtain information from the same physical device. A harder problem arises when observations made by two different physical devices are dependent on each others state; such as the problem of hand-eye coordination for example. A simple example of this is discussed in [Hager 86] and developed in Chapter 5.

3.8   Summary

If we are to understand how to fully utilize a sensors capability, we must develop a model of sensor behavior. The purpose of a sensor model is to provide a quantitative description of the ability of a sensor to extract geometric descriptions of the environment. We have developed a model of sensor capabilities in terms of a team information structure. This structure provides a probabilistic description of the information provided by a sensor as a function of the available prior information, the state of the sensor itself, and the observations or decisions made by other sensors in the multi-sensor system. We have shown how this model can be used to describe the transformation of information between different sensors, how observations depend on the location of the sensor, and how observations from other sensors can be used to guide the acquisition of information. This team-model of sensor performance will be used extensively in Chapter 5 for developing strategies of coordination and control in multi-sensor systems.

Chapter 4

INTEGRATING SENSOR OBSERVATIONS

4.1 Introduction

We will consider a multi-sensor system taking observations of the robot environment. Our goal is to integrate the partial, noisy and errorful observations provided by the sensors into a robust, consistent, consensus estimate of the state of the environment. We want any integration mechanism that we produce to be readily extensible to any number of new sensors and cues, capable of utilizing partial information, and robust to failures. We maintain a working hypothesis that by employing a large number of diverse redundant sensors in place of a smaller number of homogeneous high resolution sensors, we can achieve through sensor fusion, any specified level of task performance. This approach avoids the problem of extracting as much information as possible from a limited number of data sets and instead seeks to extract more global information from diverse, partially redundant sources. System designs based on these principles enhance reliability, increase overall system robustness and allow the exploitation of parallel processing to achieve more rapid data analysis.

The basic problem in multi-sensor systems is to integrate a sequence of observations from a number of different sensors into a single best-estimate of the state of the environment. The general estimation problem is well known and many powerful techniques exist for its solution under various conditions. The general form of these estimation techniques is to take a homogeneous sequence of observations and combine them with respect to some measure of performance to provide a single best-estimate of the environment state together with a measure of confidence in this decision. Estimation procedures rely heavily on an implicit understanding of the underlying observation characteristics.

A robot system uses notably diverse sensors which will often supply only sparse observations that can not be accurately described. In most circumstances we will have prior information about the robots environment which is often difficult to specify. In these cases we must be particularly careful to develop decision procedures which are sympathetic to the actual state of affairs encountered in the operation of the system. We can identify a number of characteristics that we should require of decision procedures that are developed with respect to poor sensor and prior modeling [Huber 81]:

- Computationally tractable. This is especially important in a robot system where we are often constrained to operate in real-time.

- Robust to both small and large deviations in observation model or prior information specification.

- Rational estimation when observations are close to expectations, regardless of sample size.

To develop decision procedures which satisfy these characteristics it is important to analyze the effects of the observation model and prior information on the resulting estimate of the state of nature.

In any given observation cycle, a robot system sensor will make a sequence of measurements $\vec{z} = \{z_1, \cdots, z_n\}$ of a set of environment states $\hat{P} = \{p_1, \cdots, p_k\}$, (edges, normals, locations, etc). Previously, we have characterized single observations, but it is also important to understand the nature of a likely sequence of observations. We consider the most important issues in multi-observation fusion to be:

- The statistical dependence between observations

- The sparsity of observations

- Possible misclassification of observations

In general, the elements of a sequence of observations $\vec{z}$ will not be statistically independent. Serious problems can arise in estimating the true state of nature if dependencies between observations are not considered. There are a number of ways that dependencies can be accounted for, although none of them are completely satisfactory. First, we could take the two extremes of complete dependency and complete independence of the observations and use them as bounds on the estimation procedure in an analogous manner to the support and belief measures in the evidential reasoning process [Dempster 68]. However, this can quickly become very complicated making consistent integration of diverse data sets difficult. Second, it may be possible to account for the dependency between observations on line, in a post-experimental fashion. This can be accomplished by either comparing observations and calculating a calibration weight [Morris 77,84], or by modeling the dependencies and estimating the parameters of the model iteratively; in a hierarchical Bayes procedure for example [Berger 85]. A third method of accounting for statistical dependencies amongst observations is to construct a prior sensor model based on a sequence of experimental observations. Kendall and Mintz [Kendall 87] have recently described a set of experiments conducted on CCD images to investigate temporal noise correlations. Their conclusions suggest that the images are reasonably pair-wise independent; although this does not guarantee ensemble independence. Much more work is needed to develop these models and to understand their effect on the observation process. In the following, we will *assume* independence in the absence of any other information.

Given the variety of possible sensor and cue observation characteristics, it is clear that we cannot expect to predict the *exact* likelihood of observing any given set of features in the environment. Chapter 3 presented an approximate model for characterizing sensor observations, described by the contaminated Gaussian distribution. We now consider a sensor taking a sequence of observations $\vec{z} = \{z_1, \cdots, z_n\}$, approximately described by the contamination model, of a feature p in the robot environment. Estimate of the feature parameter vector, based on the sample mean of the observation sequence, is inappropriate in this case because of it's sensitivity to outlying or spurious observations. Normally, a robust estimator for this type of distribution is the alpha-trimmed mean or the sample median. However, most robot sensors supply only very sparse, vector measurements of different features and so the median filter may not provide any useful results. As an alternative, suppose we have some prior information about a set of features $P = \{p_1, \cdots, p_k\}$ (edges for example) in the environment, and we take a sequence of observations $\vec{z}$ of these features which we want to use to improve our knowledge of the environment. We need to find which observations are associated with which features and which observations are spurious. To do this we must cluster the observations to our prior information in the following sense: Compare each z_i with each p_j, associating each observation with it's nearest (usually the p_j with the smallest $|p_j - z_i|$) prior feature. By applying some threshold to this comparison, we can reject observations which do not seem to correspond to any prior information, and associate each of the remaining z_i with an appropriate p_j. With no prior information, we can use a similar mechanism to compare pairs of observations z_i, z_j and cluster them into groups. In the case of an unknown environment, this will involve a general dialogue between sensor systems, generating and validating hypothesis. This is discussed further in Chapter 5. We will restrict our interest to the case of a partially known environment. This provides prior information of the observed features and permits the development of clustering procedures for contaminated observations.

When observations have been taken and clustered, we must integrate the result in to the robot system's world model. We have previously described a model of the environment which represents uncertain locations and features as probability density functions on their associated parameter vectors. These uncertain geometric objects are related to each other in a topological network. When integrating observations in to the robot's world model, it is not sufficient to just change the estimate of the observed feature, we must also change our estimate of other features so that the consistency of the world model is maintained. This requires us to propagate information through the network of object relations, preserving the topology of the world model and providing a consistent estimate of the state of the environment. The principle of maintaining consistency must be applied to all uncertain descriptions; when we update a description we must also revise our estimates of other related descriptions. Maintaining consistency also has another important consequence in that it provides a mechanism for propagating information in the world model, improving our knowledge about objects that may never even have been directly observed. The propagation of information is a fundamental concept and is a direct consequence of our desire to maintain the consistency of the robots world model.

In this Chapter, we consider the integration of disparate

sensor observations into a prior world model. Section 4.2 develops the case of a single sensor system taking observations of a single type of uncertain geometric object. Elements of Bayesian decision theory are reviewed and used to develop algorithms for clustering and filtering sparse sensor observations. Section 4.3 introduces the basic problem of integrating observations into a constrained world model. A method of describing implicit constraints and using these to propagate information is developed. Section 4.4 uses this basic constraint model to explain the problem of applying consistent updates to the location of objects in a robots world model. This integration process directly updates the constrained topological network representing the environment, and thus automatically maintains the consistency of the world model. Particular attention is given to the network of object centroids. This is because, in a partially known environment, the objects themselves are assumed known, it is just their location which is uncertain. Thus given a feature observation, we can use it to directly update the object centroid, we then need only consider consistency between uncertain object locations. Section 4.5 describes how feature observations can be used to provide partial updates of centroid locations.

The more general problem of using disparate information sources in the absence of a prior world model is considered together with the coordination and control of sensing devices in Chapter 5.

4.2 Decision Models and Information Fusion

This Section is concerned with integrating multiple observations from a single sensor. The ideas developed are based on Bayesian decision theory, using the probability density function representation for features. One of the advantages of using a distribution function to describe a feature is that it is essentially a dimensionless scalar quantity. This allows the techniques developed here to be applied to all types of features.

Information supplied by a sensor is noisy, partial and often spurious. The decision procedures developed in this section are designed to reject spurious information from sparse data sets and provide estimates that are robust to possible sensor errors. The clustering and filtering operations are able to use partial information and supply robust, computationally efficient estimates of observed features. Sections 4.2.1 and 4.2.2 will review some important elements of Bayesian decision theory and robust decision procedures. These two sections are primarily intended to introduce terminology and notation for later Chapters, and can be omitted without loss of continuity. Section 4.2.3 introduces the problem of homogeneous data fusion, and explains why direct application of standard filtering procedures is inappropriate in our situation. Section 4.2.4 develops a clustering procedure for sparse data sets, robust to deviations in model and spurious information, and capable of being extended to diverse observation sequences.

4.2.1   Decision Theory Preliminaries

We define actions a over an action space $\mathcal{A}$, the state of nature θ over a space of possible states of nature Θ and observations $\mathbf{x}$ over a sample space $\mathcal{H}$. A decision rule $\delta \in \mathcal{D}$ is a mapping $\delta: \mathcal{H} \longrightarrow \mathcal{A}$ from sample to action space such that $\forall \mathbf{x} \in \mathcal{H}, \delta(\mathbf{x}) \mapsto a \in \mathcal{A}$. For an estimation problem, the action space $\mathcal{A}$ is usually the same as the space of possible states of nature Θ: Our action is to choose an estimate $\hat{\theta} \in \Theta$. Prior information is represented as a prior density function $\pi(\theta)$ defined on Θ. With the act of making a decision we associate a loss function $L(\theta, a)$ which assigns a loss for taking an action

a when the true state of nature is θ. We define a risk function

$$R(\theta,\delta) = E_\theta^x\left[L(\theta,\delta(\mathbf{x}))\right] \tag{4.1}$$

as the expected loss (over $\mathbf{x}$) incurred by using the decision rule δ when θ is the true state of nature. We wish to find a decision rule δ such that $R(\theta,\delta)$ is as small as possible. We say a decision rule δ_1 is R-better than δ_2 if $R(\theta,\delta_1) \leq R(\theta,\delta_2)$ for all $\theta \in \Theta$ with strict inequality for at least one θ. δ is admissible if there exists no R-better decision rule. We say that a rule δ_1 is preferred to a decision rule δ_2 if $E^\pi[R(\theta,\delta_1)] < E^\pi[R(\theta,\delta_2)]$. That is we require a decision rule that minimizes

$$r(\pi,\delta) = E^\pi\left[R(\theta,\delta)\right] = \int_\Theta R(\theta,\delta)\pi(\theta)d\theta \tag{4.2}$$

The quantity $r(\pi,\delta)$ denotes the average of the risk R weighted by the prior information $\pi(\cdot)$. $r(\pi,\delta)$ is called the Bayes risk of δ, any rule $\delta^\pi \in D$ that minimizes $r(\pi,\delta)$ is called the Bayes rule, and the resulting risk $r(\pi)$ is called the minimum average risk. If $\pi(\theta)$ accurately represented our belief about θ, then δ^π would be a rational (some would say the only rational) decision rule to choose. Indeed the fundamental theorem in statistical decision theory states that any rational method of choosing amongst decision rules corresponds to the ordering induced by $r(\pi,\cdot)$ on $\delta \in D$ for some π.

To compare decision procedures we need to obtain an expression for the Bayes risk in terms of the prior information $\pi(\theta)$, loss $L(\theta,\cdot)$ and observation model $f(\mathbf{x}\mid\theta)$. Equation 4.2 can be written in the Bayes normal form as

$$r(\pi,\delta) = \int_\Theta R(\theta,\delta)\pi(\theta)d\theta = \int_\Theta\int_{\mathcal{X}} L(\theta,\delta(\mathbf{x}))f(\mathbf{x}\mid\theta)\pi(\theta)d\mathbf{x}d\theta \tag{4.3}$$

Provided L is bounded we can interchange integrals:

$$r(\pi,\delta) = \int_{\mathcal{X}}\int_\Theta L(\theta,\delta(\mathbf{x}))f(\mathbf{x}\mid\theta)\pi(\theta)d\theta\,d\mathbf{x} \tag{4.4}$$

To minimize $r(\pi, \delta)$, the decision rule $\delta(\mathbf{x})$ should be chosen to minimize

$$\int_{\Theta} L(\theta, \delta(\mathbf{x})) f(\mathbf{x} \mid \theta) \pi(\theta) d\theta \qquad (4.5)$$

for each $\mathbf{x} \in \mathcal{X}$. This is called the extensive form of Bayesian analysis. If θ and $\mathbf{x}$ have joint (subjective) density $h(\mathbf{x}, \theta) = \pi(\theta) f(\mathbf{x} \mid \theta)$ and $\mathbf{x}$ has marginal density $m(\mathbf{x}) = \int_{\Theta} f(\mathbf{x} \mid \theta) \pi(\theta) d\theta$, $m(\mathbf{x}) \neq 0$, then the posterior density of θ having observed $\mathbf{x}$ is given by:

$$\pi(\theta \mid \mathbf{x}) = \frac{h(\mathbf{x}, \theta)}{m(\mathbf{x})} = \frac{\pi(\theta) f(\mathbf{x} \mid \theta)}{m(\mathbf{x})} \qquad (4.6)$$

Hence the action $a = \delta(\mathbf{x})$ that minimizes Equation 4.5 for each $\mathbf{x}$ will also minimize

$$[m(\mathbf{x})]^{-1} \int_{\Theta} L(\theta, a) f(\mathbf{x} \mid \theta) \pi(\theta) d\theta = \int_{\Theta} L(\theta, a) \pi(\theta \mid \mathbf{x}) d\theta \qquad (4.7)$$

Thus a Bayes rule that minimizes the extensive form of the Bayes risk will also minimize the posterior expected loss.

The likelihood function is defined as $l(\theta) = f(\mathbf{x} \mid \theta)$. The likelihood principle (derived from the fundamental propositions of sufficiency and conditionality) states that $l(\theta)$ contains all the information that we need to make an estimate of θ. A maximum likelihood estimator attempts to find the value $\hat{\theta}$ that makes $l(\cdot)$ a maximum. This is a post-experimental philosophy (note that $R(\theta, \delta)$ clearly violates the likelihood principle). The maximum a posteriori estimator finds the value $\hat{\theta}$ which maximizes $\pi(\cdot \mid \mathbf{x})$. If $\pi(\theta)$ is constant (non-informative) on Θ then for a given x, the maximum of $\pi(\theta \mid \mathbf{x})$ is coincident with the maximum of $l(\theta)$. If $\pi(\theta \mid \mathbf{x})$ is unimodal and $L(\theta, a)$ is location invariant (unbiased) with respect to θ, then the maximum a posteriori estimator will be a Bayes rule.

4.2.2 Robust Decision Procedures

We will use an approximate model to describe sensor observations. The decision procedures used must therefore be robust to possible misspecification of the model $f(\cdot \mid \theta)$. In addition it is rarely (if ever) the case that we can completely specify our prior beliefs in terms of a prior distribution and so we must also be careful about possible misspecification of $\pi(\theta)$. A complete discussion of these issues is beyond the scope of this work, we will restrict ourselves to a general consideration of principles and a heuristic development of robust characteristics.

Any decision rule δ which minimize Equation 4.6 is dependent on the prior distribution $\pi(\theta)$ assumed for the unknown parameter θ. Misspecification of π will obviously affect the decision δ which is made. One way to account for possible misspecification is to consider a class Γ of plausible prior distributions π_i and obtain Bayes rules δ^{π_i} that result. These decision rules can then be used developed decision procedures which have minimum risk over all $\pi_i \in \Gamma$. This is the so-called Γ-minimax philosophy. If Γ were the class of all possible prior distributions, we would obtain a minimax decision rule. Minimax decision rules will be robust if we consider all possible outcomes, but for our purposes they have a number of undesirable characteristics; firstly such minimax rules are usually difficult to obtain, secondly they ignore any prior information that we might have, and thirdly they will in general result in less efficient algorithms if we turn out to be reasonably accurate about our model specification. A Γ-minimax procedure *is* appropriate in situations where the class of possible prior distributions is small and describes an envelope of observed sensor models.

The contaminated Gaussian sensor model is intended to approximate a sensor that normally takes observations which are relatively accurate and close to the assumed environment model, but which may provide occasional spurious or misclassified observations. This model is not intend to describe the exact character of spurious observations, rather it should force any resulting decision rule to be robust to possible misspecification. That is, we want to construct a decision rule that uses observations which correspond well to our prior model, but rejects measurements that are spurious or in error Our philosophy will be to compare observations with prior information, and if they are "close enough" to integrate the observation as if it were Gaussian, using a Bayes maximum à posteriori estimator. In this case we hope to directly remove any outlying or spurious data sets. By comparing observations and prior information before fusion, the integration process will only ever occur when we are "close" to the centroid of our prior distribution. By avoiding the outlying tails of the prior distribution we will improve our robustness to possible misspecification of outlying prior information. In this case we can choose the prior distribution to be modeled as Gaussian with sufficiently large variance. Choosing the prior distribution as a statistically pure Gaussian may be validated in a number of ways (the subject of posterior robustness is discussed at length in [Berger 85]). Firstly, as we increase the number of measurements of a particular state, the effect of any prior information on a best estimate decision becomes less important. That is, when the sample information is much greater than the prior information, it seems only natural that the prior information will have little effect on the Bayes action. This phenomenon is known as the *principle of stable estimation*. Secondly, it can also be shown, that for a decision rule δ corresponding to the mean of the posterior distribution, the quadratic (symmetric) loss, (Q a positive definite $p \times p$ matrix) $L(\delta, \theta) = (\delta - \theta)^T Q (\delta - \theta)$ is remarkably Bayes-risk robust with respect to various priors, especially when $p \geq 3$. As we continue to take independent observations, application of the central limit theorem in this context further improves the assumption of the Gaussian nature of the posterior distribution.

When the posterior distribution for a state has been calculated from the available sensor observations, we must choose a loss function on which to determine the Bayes rule for the best estimate $\hat{\theta}$ of the observed state. The choice of loss function is of great importance, it greatly affects the character of solutions to estimation problems. In the case when the distribution of θ is Gaussian, choice of an unbiased loss function (of the form $L(|\theta - a|)$, L monotone) will always result in any median of $\pi(\theta \mid x)$ as our Bayes estimator. Indeed the estimator is insensitive to the precise form of this loss providing that it is unbiased. However, using an unbiased loss is not always the best policy, for example in a compliant motion task, we would prefer to underestimate the distance to a hole, to allow for fine motion planning. We can however update the world model with the consensus view of the posterior mean and variance, and from these derive any biased best estimate that particular tasks may require.

4.2.3   Observation Fusion

We consider a sequence of observations $\vec{z} = \{z_1, \cdots, z_n\}$ of a state of the environment (a feature parameterized by Equation 2.1), which are assumed to derive from a sensor modeled by a contaminated Gaussian density, so that for the ith observation:

$$f_i(z_i, \mid p) = \left[(1 - \epsilon)N(p, \Lambda_i^1) + \epsilon N(p, \Lambda_i^2)\right] \qquad (4.8)$$

with $0.05 < \epsilon < 0.1$ and $\Lambda_i^2 \gg \Lambda_i^1$. Using Equation 4.6, we wish to integrate the observations z_i with our prior information and form a consensus best estimate of the state of the environment $\hat{p} \in P$.

We consider first the simpler case of taking observations in additive Gaussian noise:

$$z_i = p_i + v_i, \qquad v_i \sim N(0, \Lambda_i) \qquad (4.9)$$

It is well known that if the prior distribution $\pi(p)$ and the conditional observation distribution $f(z \mid p)$ are modeled as independent Gaussian random vectors $p \sim N(\hat{p}, \Lambda_0)$ and $z_1 \sim N(\hat{p}, \Lambda_1)$ respectively, then the posterior distribution $\pi(p \mid z)$

after taking a single observation z_1 can be found from Equation 4.6 and is also jointly Gaussian with mean vector

$$\hat{p}' = \left[\Lambda_0^{-1} + \Lambda_1^{-1}\right]^{-1} \left[\Lambda_1^{-1} z_1 + \Lambda_0^{-1} \hat{p}\right] \tag{4.10}$$

and variance-covariance matrix;

$$\Lambda' = \left[\Lambda_0^{-1} + \Lambda_1^{-1}\right]^{-1} \tag{4.11}$$

Consider now n independent observations z_i $(i = 1, \ldots, n)$ of p each sampled from a different $f_i(z_i \mid p)$ and each considered Gaussian as $N(p, \Lambda_i)$. Defining $z_0 \equiv \hat{p}$, it can be shown by induction on Equations 4.10 and 4.11 that the posterior distribution $\pi(p \mid z_1, \cdots, z_n)$ (our belief about p after making the observations $z_1, \cdots, z_n$) is again jointly Gaussian with mean:

$$\hat{p}' = \left[\sum_{i=0}^{n} \Lambda_i^{-1}\right]^{-1} \left[\sum_{i=0}^{n} \Lambda_i^{-1} z_i\right] \tag{4.12}$$

and variance-covariance matrix

$$\Lambda' = \left[\sum_{i=0}^{n} \Lambda_i^{-1}\right]^{-1} \tag{4.13}$$

whence the maximum à posteriori estimate of p is the mean $\hat{p}'$. This estimator, a discrete Kalman filter, is attractive because of its computational efficiency. If the assumption that the z_i are jointly Gaussian is indeed valid, then this filter will be optimal in a minimum variance sense. If the observations are not Gaussian, this filter will still be the optimum *linear* filter for symmetric loss functions [Gelb 74].

In general it is unrealistic to model the $f_i(z_i \mid p)$ as jointly Gaussian observations because of the possible flattening of the distribution tails and the inclusion of spurious and misclassified observations. However in keeping with our goal of providing a decision procedure that is efficient when close to the assumed model and otherwise robust to deviations, we shall form a policy that classifies observations, rejects outliers and allows the remaining observations to be considered as deriving from a jointly Gaussian model, so that the computationally efficient estimator of Equations 4.12 and 4.13 may be used to combine the remaining observations.

We consider the $f_i(z_i \mid p)$ as modeled by contaminated Gaussian distributions. This should be interpreted as a belief that the required observations are considered Gaussian but are embedded in further contaminations that include other spurious data sets. It is well known that the contaminated Gaussian model is *not* jointly Gaussian, and hence it is not theoretically (or intuitively) possible to apply Equations 4.12 and 4.13 to the unprocessed observations. The "traditional" approach in this situation is to use approximations based on a large sample assumption of the ordered statistics: If we assume the observations z_i are scalar, we can order them by size and find the median of the sample set. It can be demonstrated that the sample median of a sample of contaminated Gaussian variates is asymptotically Gaussian [Mintz 86]. Thus if we remove the outlying α fraction of the sample, the remaining observations can be approximated as being derived from a pure Gaussian sample. The mean of this reduced sample is called the α-trimmed mean.

Although estimators based on the sample median are known to be robust, their use in conjunction with the sensor systems that we are considering has a number of problems. In general we will be making observations of vector quantities which do not admit an obvious sample ordering or sample median[1]. Of greater importance is the inability of these techniques to address the possibility of misclassification of observations, particularly in the case where the z_i are not identically distributed

---

[1]However, treating each element of an observed vector individually, it may be possible to overcome this objection at the cost of added complexity.

and do not represent just a single state of nature, but a *set* of possibly unknown states. There is no clear way that ordered statistics techniques can be directly applied to this problem. In addition, the sensors of a robot system rarely supply dense data in sufficient quantities to justify the large sample assumptions required of the ordered statistics techniques (a laser range finder is one exception); for example, a vision system will often supply but a single measurement of an edge, applying an α-trim to such a small sample is meaningless.

We must develop a decision procedure capable of using sparse vector observations of many different features. In the following we will seek a methodology that incrementally classifies observations in to clusters that can be considered as jointly Gaussian, rejects outlying measurements and is robust to deviations in the sensor model.

4.2.4 Sparse Data Fusion

We again assume a sequence of observations $\vec{z} = \{z_1, \cdots, z_n\}$ derived from a sensor modeled by a contaminated Gaussian density. Idealy the observations will be associated with the measurement of a single state of nature, assuming prior clustering to have already been performed. However it is clear that some z_i may be spurious or poor data sets and considerable misclassification may well have occurred. For this reason we consider the z_i as observations on a set of possible features in the environment $P = \{p_1, \cdots, p_k\}$. We wish to find a method of clustering the z_i so that the outlying observations can be removed from consideration and the remaining z_i assigned to appropriate $p_j \in P$. Removal of outliers enables the remaining z_i to be approximated as coming from a set of pure Gaussian observations (one in each possible $p_j \in P$) which can then be integrated using Equations 4.12 and 4.13. If P is not known à priori then to cluster the observations we must do a pair-wise comparison of the z_i, z_j in the following sense: If z_j is assumed to derive from a Gaussian distribution, can z_i be explained by the distribution on z_j, and similarly can z_j be explained by the distribution on z_i, further if agreement can be reached, what is the consensus value $\hat{p}$ of z_i and z_j? Clearly any such comparison will take the form;

$$F(|z_i - z_j|) \leq T_{ij} \tag{4.14}$$

where F is some monotone increasing function and T_{ij} is some positive threshold. There are a number of ways that this criterion can be reached, most obviously by some kind of hypothesis test. We choose to derive this clustering process in terms of the preference ordering induced by the observation model $f(\cdot \mid p)$ on the observation set $\vec{z}$. The reason for this choice is that clustering based on the distribution function f is essentially dimensionless and as a consequence, can be used to cluster geometrically disparate information. The extension to this case will be considered in Chapter 5.

A distribution function is essentially a measure of observation likelyhood; the higher the probability of observation, the more likely an observation is to occur. This likelyhood can be considered as a measure of preference on observations; we would prefer more likely things to be observed. Thus the mode of a distribution is the most preferred observation value, with decreasing preference as we move away from the mode. To provide a complete ordering on observations, these preferences must form a convex set; so that the preference measure is unimodal. In terms of density functions this requires that the likelyhood function be convex over the area of interest[2]. Thus the posterior distribution function $f(p \mid z_i)$ associated with the observation z_i can also be considered as the preference ordering induced by z_i on any resulting estimate $\hat{p}$. In this framework,

---

[2]This is equivalent to the well-known Blackwell-Girshick theorem [Blackwell 54]. This states that non-randomized decision rules are only optimal in cases where the set of observation utilities is convex. In all other cases, randomized procedures produce higher utilities.

two observations z_i and z_j can only form a joint estimate if the posterior distribution $f(p \mid z_i, z_j)$ calculated from Bayes rule is unimodal.

With z_i observed and fixed, we need to find the range of p over which f_i is convex. A function f_i will be convex if and only if its matrix of second order derivatives is positive semi-definite. Assuming that the observations are modeled as Gaussian; $f_i(z_i \mid p) \sim N(\hat{p}, \Lambda_i)$, differentiating f_i gives

$$\frac{\partial^2 f_i}{\partial p^2} = \Lambda_i^{-1} \left[1 - (p - z_i)^T \Lambda_i^{-1} (p - z_i) \right] \cdot N(\hat{p}, \Lambda_i) \quad (4.15)$$

For this to be positive definite and hence f_i to be convex for a fixed z_i, we are required to find an estimate p which satisfies;

$$(p - z_i)^T \Lambda_i^{-1} (p - z_i) \le 1 \quad (4.16)$$

for all $i = 1, \cdots, n$ observations.

Consider any two observations z_i and z_j. They can form a joint estimate if we can find a p that satisfies Equation 4.16 for both z_i and z_j. As the left hand side of Equation 4.16 must always be positive, this is equivalent to finding a p which satisfies

$$\frac{1}{2}(p - z_i)^T \Lambda_i^{-1} (p - z_i) + \frac{1}{2}(p - z_j)^T \Lambda_j^{-1} (p - z_j) \le 1 \quad (4.17)$$

The value of p which makes the left hand side of this equation a minimum (and which is also the estimate when it exists) is given by the weighted average

$$p = \left(\Lambda_i^{-1} + \Lambda_j^{-1} \right)^{-1} \left(\Lambda_i^{-1} z_i + \Lambda_j^{-1} z_j \right) \quad (4.18)$$

Substituting this in to Equation 4.17 gives

$$\frac{1}{2}(z_i - z_j)^T (\Lambda_i + \Lambda_j)^{-1} (z_i - z_j) \le 1 \quad (4.19)$$

We will say that z_i and z_j admit a joint maximum likelihood estimate $\hat{p}$ if they satisfy Equation 4.19. If the estimate $\hat{p}$ exists it will correspond to the unique maximum of $f(p \mid z_i, z_j)$ given by Equation 4.18.

Equation 4.19 can be used to compare observations and cluster them into groups which may be considered as Gaussian. In most real situations, it is unlikely that we will know the variance-covariance matrices exactly. In this case, any estimates of the Λ_i act like thresholds; the larger the Λ_i that is used, the more disagreement will be tolerated. This is in agreement with the suggestion presented previously that any comparison of observations will be of the form of Equation 4.14.

These observation clusters can be built recursively, making use of local geometric information to separate cluster candidates. The largest clusters can then be used to provide world model updates, the smaller clusters can be rejected as outlying observations. This clustering process allows sparse observations to be integrated in an efficient and robust manner. Use of the information matrix in Equation 4.19 allows partial information to be compared and utilized consistently. The dimensionless form of the clustering process allows any type of parameterized feature to be integrated. Chapter 5 will also show that geometrically different observations can be compared using this method, providing a mechanism to cluster disparate sensor opinions.

4.3   Integrating Observations with Constraints

Our model of the world and sensor systems can be abstractly described as a constrained (topological) network. The object features, sensors and coordinate frames correspond to nodes, and the uncertain relations between these nodes correspond to the arcs of this network. We label the arcs with random variables (vectors) e_i each with an associated prior probability density function $\pi_i(e_i)$ representing the (possibly unknown) uncertainty in relations between nodes. An observation taken in the robot environment corresponds to a new measurement z_i of an arc e_i in the network abstraction. The fusion of a new set of sensor observations into a prior network of relations cannot be accomplished independently for each relation arc. This is because the network is constrained, and so updating a single relation (arc) requires that other relations be changed so that consistency can be maintained. Our goal is to accomplish this fusion of observations and prior constraints in a consistent and rational manner.

We consider first the case where the arc labels e_i and the observations z_i are scalar. It is assumed that the observations have previously been clustered and filtered as described in the previous section. A method is then developed for the fusion of these observations with a random network representing geometric constraints in the world model of the robot system. Section 4.4 uses the methodology developed in the scalar case to solve the problem of making consistent changes in object centroid locations, resulting from an observation of object features.

4.3.1 Updating a Constrained Random Network

Continuing our Bayesian philosophy, we consider taking observations $z = (z_1, \cdots, z_k)^T$ of the network arcs $e = (e_1, \cdots, e_m)^T$ (relations in the environment), modeled by a sample distribution $f(z \mid e)$. Suppose that this network is considered as prior information so that the posterior distribution (our belief about the network after making the observations z) is given by Bayes rule:

$$\pi(e \mid z) = \frac{f(z \mid e)\pi(e)}{m(z)} \qquad (4.20)$$

Then using a maximum à posteriori estimator, our new arc labeling estimates will be the $\hat{e}$ that maximize $\pi(e \mid z)$.

Unfortunately Result 3.2 of Chapter 2 shows that Λ_e must always be singular. Consequently $\pi(e)$ and $\pi(e \mid \cdot)$ are degenerate and Equation 4.20 makes no sense. This situation arises because the network constraints on e are represented internally and implicit in $\pi(e)$. Thus external Bayesianity of the network is guaranteed for any consistent internal estimates $\hat{e}$ satisfying $C\hat{e} = 0$. Clearly Equation 4.20 is not directly useful for finding a rational update policy.

An alternative is to require that each relation (arc) be updated by the observations in an independently Bayesian fashion while continuing to satisfy these network constraints. This internal Bayesian rationality coupled with the externally consistency requires that the *distributions* in the arc labels e be updated as:

$$\pi_i(e_i \mid z) = \frac{f(z \mid e_i)\pi_i(e_i)}{m(z)} \qquad (4.21)$$

The maximum à posteriori estimate of the arc labelings is obtained by finding the vector $\hat{e} = [\hat{e}_1, \cdots, \hat{e}_m]^T$ that maximizes Equation 4.21 for each $i = 1, \cdots, m$ subject to the constraints

$$C\hat{e} = 0 \quad \text{and} \quad C\Lambda_e C^T = 0 \qquad (4.22)$$

In the following we will assume that prior clustering of the observations has already occurred so that for each arc e_i, there is at most one observation z_i, which admits a Bayesian consensus with e_i.

Initially consider the case where each observation z_i is independent and modeled as a pure Gaussian distribution $f(z_i \mid \cdot) \sim N(e_i, \gamma_i)$ (this constraint will be relaxed later). Further assume that the prior marginal distributions on the network arcs $\pi_i(\cdot)$ are Gaussian as $N(e_i, \sigma_i)$ so that the posterior distributions of the arc labels are given by

$$\pi_i(e'_i \mid z_i) = \frac{1}{\sqrt{2\pi}\bar{\sigma}_i} \exp\left[-\frac{(e'_i - \bar{e}_i)^2}{2\bar{\sigma}_i^2}\right] \tag{4.23}$$

where

$$\bar{e}_i = \left(\frac{e_i}{\sigma_i^2} + \frac{z_i}{\gamma_i^2}\right)\left(\frac{1}{\sigma_i^2} + \frac{1}{\gamma_i^2}\right)^{-1} \tag{4.24}$$

and

$$\bar{\sigma}_i^2 = \left(\frac{1}{\sigma_i^2} + \frac{1}{\gamma_i^2}\right)^{-1} \tag{4.25}$$

Equations 4.15 and 4.16 show that the value of e'_i that maximizes Equation 4.23 also minimizes the quadratic defined by

$$L_i(e'_i) = \frac{(e'_i - \bar{e}_i)^2}{2\bar{\sigma}_i^2} = \frac{1}{2}\frac{(e'_i - e_i)^2}{\sigma_i^2} + \frac{1}{2}\frac{(e'_i - z_i)^2}{\gamma_i^2} \tag{4.26}$$

This allows us to restate the problem of finding a consistent fusion of observations into a network of prior constraints, as finding the labels $\hat{e}$ that minimize $L_i(\cdot)$ for $i = 1, \cdots, m$ subject to $C\hat{e} = 0$.

Define Σ_e $(m \times m)$ to be the diagonal matrix of arc label variances σ_i^2 $(i = 1, \cdots, m)$ and define Λ_z to be the (nonsingular, not necessarily diagonal) variance-covariance matrix of the observation vector $z = [z_1, \cdots, z_k]$, $(k \leq m)$. Let N_z be a matrix of dimension $k \times m$ composed of a $k \times k$ identity matrix I, padded by zeros so that $(N_z e - z)$ has a sensible interpretation. Define

$$L(e') = \sum_{i=1}^{m} L_i(e'_i) \tag{4.27}$$

It is clear that because the L_i are positive quadratics, minimizing each L_i individually with respect to the appropriate e'_i

is equivalent to minimizing $L(\cdot)$ with respect to the vector e'. Then Equation 4.26 can be summarized by minimizing:

$$L(e') = \tfrac{1}{2}(e' - e)^T \Sigma_e^{-1} (e' - e)$$
$$+ \tfrac{1}{2}(N_z e' - z)^T \Lambda_z^{-1} (N_z e' - z) \qquad (4.28)$$

subject to

$$G(e') = Ce' = 0 \qquad (4.29)$$

We will minimize by Lagrange multipliers [Dorny 74]. The constrained minimum of L is given by the vector $\hat{e}$ that satisfies;

$$\frac{\partial L}{\partial e'} = \frac{\partial G}{\partial e'}\lambda \qquad (4.30)$$

where λ ($r' \times 1$) are the unknown Lagrange multipliers. Differentiating L and G yields:

$$\frac{\partial L}{\partial e'} = \Sigma_e^{-1}(e' - e) + N_z^T(N_z e' - z)$$
$$\frac{\partial G}{\partial e'} = C^T \qquad (4.31)$$

Substituting Equation 4.31 into Equation 4.30 gives:

$$\Sigma_e^{-1}(\hat{e} - e) + \Lambda_z^T\left(N_z^{-1}\hat{e} - z\right) = C^T\lambda \qquad (4.32)$$

Multiplying by the incidence matrix M and using the fact that $MC^T = 0$ results in:

$$M\left[\Sigma_e^{-1}(\hat{e} - e) + \Lambda_z^{-1}(N_z\hat{e} - z)\right] = MC^T\lambda = 0 \qquad (4.33)$$

Rearranging in terms of update and observation gives:

$$M\left[\Sigma_e^{-1} + N_z^T\Lambda_z^{-1}N_z\right](\hat{e} - e) = MN_z^T\Lambda_z^{-1}(z - N_z e) \qquad (4.34)$$

This gives the required changes in arc labels ($\hat{e} - e$) in terms of the difference between observations and prior information

$(\mathbf{z} - \mathbf{N}_z\mathbf{e})$. The resulting $\hat{\mathbf{e}}$ satisfy the required geometric constraints, maintain external consistency and provide internal Bayesianty.

Equation 4.34 represents n equations in m unknowns, we must find an independent set of $n-1$ arc labels to update from this equation. Recall that we can choose an arbitrary node and connect it to all other nodes by arcs labeled x_i, that satisfy $\mathbf{e} = \mathbf{M}^T\mathbf{x}$. Clearly the $\mathbf{x}$ form an independent set of labels (the network cannot be specified without all of them). Choose a node in the network to which is connected an arc e_j that has been observed as z_j, then the labels $\mathbf{x}^j$ from this node form an independent *subset* of $\mathbf{e}$ that satisfy $\mathbf{e} = \mathbf{M}^T\mathbf{x}^j$ and have $x_j^j = 0$. The update to these independent labels can then be found from;

$$\mathbf{M}\left[\mathbf{\Sigma}_e^{-1} + \mathbf{N}_z^T\mathbf{\Lambda}_z^{-1}\mathbf{N}_z\right]\mathbf{M}^T\left(\hat{\mathbf{x}}^j - \mathbf{x}^j\right) = \mathbf{M}\mathbf{N}_z^T\mathbf{\Lambda}_z^{-1}\left(\mathbf{z} - \mathbf{N}_z\mathbf{e}\right)$$

$$(4.35)$$

with $\hat{\mathbf{e}} - \mathbf{e} = \mathbf{M}^T\left(\hat{\mathbf{x}}^j - \mathbf{x}^j\right)$ From Result 3.2 of Chapter 2, the matrix

$$\mathbf{M}\left[\mathbf{\Sigma}_e^{-1} + \mathbf{N}_z^T\mathbf{\Lambda}_z^{-1}\mathbf{N}_z\right]\mathbf{M}^T \qquad (4.36)$$

is always singular, hence one of the elements of $(\hat{\mathbf{x}}^j - \mathbf{x}^j)$ is arbitrary. Here we have fixed $\hat{x}_j^j - x_j^j = 0$. The remaining labels can be found by Gaussian elimination. The new updated information matrix of arc labels is just given be Equation 4.36.

4.3.2 The Three-Node Example

To fix these ideas, it is useful to consider a simple example. Assume a prior network of three nodes (objects) and three edges (relations), with

$$\mathbf{M} = \begin{bmatrix} 1 & 0 & -1 \\ -1 & 1 & 0 \\ 0 & -1 & 1 \end{bmatrix}, \mathbf{C} = [1 \quad 1 \quad 1], \mathbf{\Sigma}_e = \begin{bmatrix} \sigma_1^2 & 0 & 0 \\ 0 & \sigma_2^2 & 0 \\ 0 & 0 & \sigma_3^2 \end{bmatrix}.$$

Consider taking an observation z of the arc e_1 with $\Lambda_z = \gamma^2$ and $N_z = [1, 0, 0]$. Define $d = (z - e_1)$, then from Equation 4.35 we have;

$$\begin{bmatrix} \frac{1}{\sigma_1^2} + \frac{1}{\sigma_3^2} + \frac{1}{\gamma^2} & -\frac{1}{\sigma_1^2} - \frac{1}{\gamma^2} & -\frac{1}{\sigma_3^2} \\ -\frac{1}{\sigma_1^2} - \frac{1}{\gamma^2} & \frac{1}{\sigma_1^2} + \frac{1}{\sigma_2^2} + \frac{1}{\gamma^2} & -\frac{1}{\sigma_2^2} \\ -\frac{1}{\sigma_3^2} & -\frac{1}{\sigma_2^2} & \frac{1}{\sigma_2^2} + \frac{1}{\sigma_3^2} \end{bmatrix} (\hat{x} - x) = \begin{bmatrix} -\frac{d}{\gamma^2} \\ \frac{d}{\gamma^2} \\ 0 \end{bmatrix}$$

Fixing $\hat{x}_1 - x_1 = 0$ results in

$$
\begin{aligned}
(\hat{x}_1 - x_1) &= 0 \\
(\hat{x}_2 - x_2) &= (\sigma_1^2 + \gamma^2)^{-1} \sigma_2^2 \left(\sigma_2^{-2} + \sigma_3^{-2} + (\sigma_1^2 + \gamma^2)^{-1} \right)^{-1} \frac{d}{\gamma^2} \\
(\hat{x}_3 - x_3) &= (\sigma_1^2 + \gamma^2)^{-1} \sigma_3^2 \left(\sigma_2^{-2} + \sigma_3^{-2} + (\sigma_1^2 + \gamma^2)^{-1} \right)^{-1} \frac{d}{\gamma^2}
\end{aligned}
$$

The $\hat{e}$ can be found from $\hat{e} = M^T(\hat{x} - x) + e$. Note the similarity of these results to the pure Bayes estimator.

4.4 Estimating Environment Changes

We will consider the case of a known but uncertain environment in which the observed objects are known, but their location is uncertain. In this case, we will observe the different features that comprise the object and use them to update the object's location. Different sensors can then communicate their observations in terms of proposed changes in object locations. To integrate geometrically dissimilar measurements, we can just integrate these local estimates of location.

To achieve this process of communication and update, we must first describe changes in object location, and show how individual locations can be updated from partial observations of different features.

4.4.1   Changes in Location

The relation between two objects (centroids) can be represented in two ways, either by a description vector or by a homogeneous transform; one representation can always be obtained from the other. We will use both because a description vector is much easier to compare with observations, whereas a homogeneous transform is considerably easier to use when transforming information around a network of relations. A description vector $\mathbf{D}$ is composed of two parts, a position vector $\mathbf{p}$ and an orientation vector $\mathbf{r}$ (in roll-pitch-yaw representation), defined by:

$$\mathbf{D} = \begin{bmatrix} \mathbf{p} \\ \mathbf{r} \end{bmatrix}; \quad \text{where} \quad \mathbf{p} = \begin{bmatrix} p_x \\ p_y \\ p_z \end{bmatrix} \quad \text{and} \quad \mathbf{r} = \begin{bmatrix} \psi \\ \theta \\ \phi \end{bmatrix} \quad (4.37)$$

Consider a change in relation between two object centroids in terms of the homogeneous transform that relates them. The difference between an initial transform $\mathbf{T}$ and an updated transform $\mathbf{T}'$ can be described by the difference between their associated descriptions vectors:

$$\mathbf{d} = \mathbf{D}' - \mathbf{D} = (\delta_x, \delta_y, \delta_z, \delta_\psi, \delta_\theta, \delta_\phi,)^T \quad (4.38)$$

If this change is small, the difference in transforms can be written in the form of a differential transform $\mathbf{\Delta}_T$ [Paul 81] (the skew symmetric second order tensor of d). So that if

$$\mathbf{T}' = \mathbf{T} + \delta\mathbf{T} = (\mathbf{\Delta}_T + \mathbf{I})\,\mathbf{T} \quad (4.39)$$

then

$$\delta\mathbf{T} = \mathbf{\Delta}_T\mathbf{T} \quad (4.40)$$

where

$$\mathbf{\Delta}_T = \begin{bmatrix} 0 & -\delta_\phi & \delta_\theta & \delta_x \\ \delta_\phi & 0 & -\delta_\psi & \delta_y \\ -\delta_\theta & \delta_\psi & 0 & \delta_z \\ 0 & 0 & 0 & 0 \end{bmatrix} \quad (4.41)$$

If ^iA_j is a transform from frame i in to a coordinate frame j, and ^id ($^i\Delta$) is a change in description vector in coordinate frame i, then the equivalent change in description vector ^jd ($^j\Delta$) in frame j can be found from

$$^j\Delta = \,^iA_j\,^i\Delta\,^iA_j^{-1} \tag{4.42}$$

4.4.2   Using Feature Observations to Update Centroid Locations

We will assume a known environment, ensuring that the relation between features and centroids is known and fixed. There are two methods by which feature observations may be updated and propagated; through the relations between features themselves or directly through the object centroid. Propagation of feature observations through the feature network itself is considerably more difficult than updating directly through the object centroid. However using the feature network is more readily extended to the cases of partially or completely unknown environments. Here we restrict ourselves to considering only propagation through centroid locations.

Consider a sensor taking observations $z \sim N(p, \Lambda_p)$ of a feature $g(x, p) = 0$ in the environment. Let T_s and T_o be homogeneous transforms describing the location of a sensor and the object centroid respectively in terms of some fixed but arbitrary coordinate frame [Figure 4.1]. Let $p = h(D)$ describe the relation between the feature parameter vector and the object centroid. Generaly h will be known although h^{-1} will usually be indeterminate. For example, given a known object we can find it's edges, but given one of the edges we can not uniquely determine the object centroid. If the location of the object centroid D is uncertain, then we can easily approximate the distribution on a feature vector p: Let $J_h = \frac{\partial h}{\partial D}$ and suppose $D \sim N(\hat{D}, \Lambda_D)$, then

$$p \sim N(\hat{p}, \Lambda_p),$$

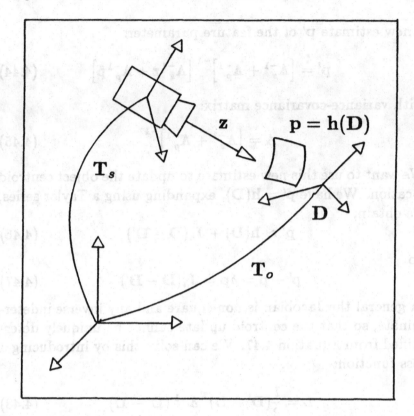

Figure 4.1: Relation Between Sensors, Features and an Object Centroid

with
$$\hat{p} = h(\hat{D}) \qquad \text{and} \Lambda_p = J_h \Lambda_D J_h^T.$$

Suppose we now take observations z of the feature vector, assumed to derive from a contamination model, then we can cluster these observations as:

$$\frac{1}{2}(z - \hat{p})^T (\Lambda_z + \Lambda_p)^{-1} (z - \hat{p}) \le 1 \qquad (4.43)$$

This associates the observations with the correct prior feature and object centroid. After filtering, these observations provide a new estimate $\mathbf{p}'$ of the feature parameter:

$$\mathbf{p}' = \left[\mathbf{\Lambda}_z^{-1} + \mathbf{\Lambda}_p^{-1}\right]^{-1}\left[\mathbf{\Lambda}_z^{-1}\mathbf{z} + \mathbf{\Lambda}_p^{-1}\hat{\mathbf{p}}\right] \qquad (4.44)$$

with variance-covariance matrix:

$$\mathbf{\Lambda} = \left[\mathbf{\Lambda}_z^{-1} + \mathbf{\Lambda}_p^{-1}\right]^{-1} \qquad (4.45)$$

We want to use this new estimate to update the object centroid location. We have $\mathbf{p}' = \mathbf{h}(\mathbf{D})$, expanding using a Taylor series, we obtain:

$$\mathbf{p}' \approx \mathbf{h}(\mathbf{D}) + \mathbf{J}_h(\mathbf{D} - \mathbf{D}') \qquad (4.46)$$

So

$$\mathbf{p}' - \mathbf{p} = \delta\mathbf{p} = \mathbf{J}_h(\mathbf{D} - \mathbf{D}') \qquad (4.47)$$

In general the Jacobian is non-square and any inverse indeterminate, so that the centroid update cannot be uniquely determined from Equation 4.47. We can solve this by introducing a loss function:

$$L = \frac{1}{2}(\mathbf{D}' - \mathbf{D})^T\mathbf{\Lambda}^{-1}(\mathbf{D}' - \mathbf{D}) \qquad (4.48)$$

The update $(\mathbf{D}' - \mathbf{D})$ is found by minimizing L with respect to $\mathbf{D}'$ subject to the constraint given by the feature update Equation 4.47. By Lagrange multipliers, we obtain:

$$(\mathbf{D}' - \mathbf{D}) = \mathbf{\Lambda}\mathbf{J}_h^T\left(\mathbf{J}_h\mathbf{\Lambda}\mathbf{J}_h^T\right)^{-1}\delta\mathbf{p} \qquad (4.49)$$

Thus, given the change in feature estimate $\delta\mathbf{p}$, we can find the change in centroid estimate $(\mathbf{D}' - \mathbf{D})$ that satisfies this new observation and provides a minimum risk change in location.

4.4.3   Logical Relations and Updates

Objects in the world are often constrained to exist in a subset of possible states, either physically or for stability reasons. For example a box will in general stand on a face, not a vertex (stability) or a desk draw can only move on its runners (physical constraint). We consider the relation *on* as an example and explain its setting in the update and propagation algorithm. This example is not intended to be exhaustive, rather indicative of procedure. The relation *on* allows us to constrain the possible relative orientations between related objects. For example, a box sitting on a desk must in general (excluding pathological exceptions) have one axis normal to the desk top, so that its orientation with respect to the desk is constrained to be one of:

$$
\begin{bmatrix} . & . & 0 \\ . & . & 0 \\ 0 & 0 & \pm1 \end{bmatrix}
\begin{bmatrix} \pm1 & 0 & 0 \\ 0 & . & . \\ 0 & . & . \end{bmatrix}
\begin{bmatrix} . & 0 & . \\ 0 & \pm1 & 0 \\ . & 0 & . \end{bmatrix}
\tag{4.50}
$$

This constraint can be represented as a "virtual" measurement of the relative description vector with no uncertainty in one degree of freedom (normal to the desk) and infinite uncertainty in the other descriptors. The Bayesian integration technique will naturally cause this zero variance descriptor to dominate the consensus, while the descriptors with infinite uncertainty will essentially be ignored.

Thus the logical relation may be considered as an observation in the manner described previously, and applied to the updating and propagation algorithms in an identical manner to normal observations.

4.5   Consistent Integration of Geometric Observations

We now have a means of providing changes in object location from a set of disparate observations made of an uncertain environment. We must be able to apply these changes to the robots world model in a consistent manner. In this section, we

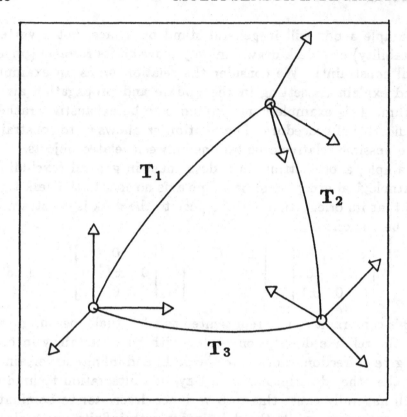

Figure 4.2: Three Objects In a Relational Loop

will describe the basic problem of maintaining consistency in a network of centroid locations and show how sensor information can be integrated into this network. This integration of sensor information with location constraints allows consistency between geometric relations to be maintaining while making full use of available sensory information.

4.5.1   The Consistency Problem

Suppose we have three objects related as in Figure 4.2, and assume that we have prior information as to the relations between them in terms of the homogeneous transforms T_1, T_2

and T_3 each with some associated uncertainty measure. Further assume that this prior network is indeed consistent so that the relation

$$T_1 T_2 T_3 = I \qquad (4.51)$$

is satisfied by the best estimates. If we now take a measurement of one of the relations using a sensor in the system, we must preserve the consistency of this network when we update the observed relation. For example, suppose we take a new "measurement" of T_1, and obtain a new consensus best estimate of this transform as T_1'. Clearly $T_1' T_2 T_3 \neq I$; the world model will no longer be internally consistent. There have been two main solutions to this problem:

1. Rather than using best estimates, we take the union of the errors of the new and the old values of the transform T_1. This is similar to the uncertainty ellipsoid method used by Brooks [Brooks 84]. The problem here is that making more sensor observations, although certain to enclose the right best estimate, will increase the uncertainty in the world model, leading one to believe that taking less sensor observations would be the best policy.

2. If we neglect the fact that internal consistency must be maintained after integrating a new observation, then relations between objects must be redefined. In this example, a new transform T_4 (say) is introduced, so that $T_1' T_4 T_2 T_3 = I$ holds. This is similar to the problem encountered with the HILARE mobile robot [Chatila 84], whereby after a number of location updates, the transform T_4 becomes significant enough for the world model to introduce a new object between T_1 and T_2 in order to satisfy the consistency constraints.

A better solution is to identify a new T_2' and T_3' so that we maintain consistency in the form:

$$T_1' T_2' T_3' = I \qquad (4.52)$$

However, with a fixed new estimate T_1', this Equation is not sufficient to uniquely determine both T_2' and T_3'. In fact *any* estimates of these transforms satisfying $T_1' = (T_2' T_3')^{-1}$ would maintain consistency, the problem is to find which estimates are best under this constraint.

We will apply the method developed Section 4.3 to this problem, and develop an algorithm which integrates observations in to a network of geometric constraints, and which is internally Bayesian and externally consistent.

4.5.2   Consistent Changes in Location

We will consider an arbitrary number of objects with their centroids connected to each other by any number of relations. This assumption gives the world model the character of a network. The problem is that if we change our best estimate of the location of a particular object, how does this affect (through consistency) all the other relations and states in the world model. We draw an analogy here with a network of objects connected to each other by springs (relations). At equilibrium the forces applied by the springs at each object exactly balance each other. If we now move one of the objects (or equivalently change the tension in one of the springs), the forces in the adjacent springs change, and thus propagate the original motion through the network. At the new equilibrium, the forces applied by the springs at each object again exactly balance. We note that the effect of moving a particular object is strongest in the local vicinity, but decreases with distance. In our world model, when we update the state of an object we must also update relations (spring tensions) and objects in its neighborhood in order that consistency (equilibrium) is obtained.

We can describe the world model as a general directed network in terms of an incidence matrix M, having a row for every object (node) and a column for every relation (edge). The consistency constraint in the case of a general network of relations is to require that the product of the transforms around each loop of relations is the identity:

$$\prod_{i \in L_j} \mathbf{T}_i = \mathbf{I} \qquad \forall L_j \in \{L \mid L \text{ is a loop in M}\} \qquad (4.53)$$

If there are r loops in the network, the consistency problem can be stated as r non-linear matrix equations in m variables. By introducing differential transforms, the consistency conditions can be approximated by a set of linear matrix equations. After an update Equation 4.53 must be satisfied by the resulting new transforms $\mathbf{T}'_j$:

$$\prod_{i \in L_j} \mathbf{T}'_i = \mathbf{I} \qquad \forall L_j \in \{L \mid L \text{ is a loop in M}\} \qquad (4.54)$$

Then from Equation 4.39

$$\mathbf{I} = \prod_{i \in L_j} \mathbf{T}'_i = \prod_{i \in L_j} (\mathbf{T}_i + \delta \mathbf{T}_i) \qquad \forall L_j \qquad (4.55)$$

Expanding, neglecting second order terms and using Equation 4.40, we obtain:

$$\prod_{i \in L_j} \mathbf{T}_i + \sum_{i \in L_j} \left(\prod_{\substack{k \in L_j \\ k < i}} \mathbf{T}_k \right) {}^i\mathbf{\Delta}_{T_i} \left(\prod_{\substack{k \in L_j \\ k > i}} \mathbf{T}_k \right) = \mathbf{I} \quad \forall L_j \quad (4.56)$$

Where the ${}^i\mathbf{\Delta}_{T_i}$ are the required differential changes in the transforms $\mathbf{T}_i$ that maintain consistency. Multiplying by Equation 4.53 and simplifying gives:

$$\sum_{i \in L_j} \left(\prod_{\substack{k \in L_j \\ k \le i}} \mathbf{T}_k \right) {}^i\mathbf{\Delta}_{T_i} \left(\prod_{\substack{k \in L_j \\ k \le i}} \mathbf{T}_k \right)^{-1} = 0 \quad \forall L_j \qquad (4.57)$$

Using Equation 4.42 we can reference each ${}^i\Delta_{T_i}$ to a fixed coordinate frame ($i = 0$ say), so that Equation 4.57 can be written as:

$$\sum_{i\in L_j} {}^0\Delta_{T_i} = 0 \qquad \forall L_j \tag{4.58}$$

Rewriting this in terms of changes in description vector, we have

$$\sum_{i\in L_j} {}^0d_i = 0 \qquad \forall L_j \tag{4.59}$$

This is a vector version of the scaler labeling problem considered earlier. Equation 4.59 states that the changes in description vector around a loop referenced to a common coordinate frame must sum to zero. As before, define **C** to be the matrix describing arc loops with 6×6 identity matrices in place of the usual 1 elements. With ${}^0d = [{}^0d_1^T, \cdots, {}^0d_m^T]^T$, Equation 4.59 can be summarized by

$$\mathbf{C}^0d = \mathbf{C}({}^0\mathbf{D}' - {}^0\mathbf{D}) = 0 \tag{4.60}$$

This represents the linearized location network constraints required to maintain consistency.

For example the relation network of Figure 4.2, can be described by

$$\mathbf{C} = [\mathbf{I}, \mathbf{I}, \mathbf{I}], \quad \text{or,} \quad \mathbf{M} = \begin{bmatrix} \mathbf{I} & 0 & -\mathbf{I} \\ -\mathbf{I} & \mathbf{I} & 0 \\ 0 & -\mathbf{I} & \mathbf{I} \end{bmatrix}.$$

With $\mathbf{T_1 T_2 T_3} = \mathbf{I}$, the consistency constraint $\mathbf{T_1' T_2' T_3'} = \mathbf{I}$ reduces to:

$$\left({}^0\mathbf{D_1'} - {}^0\mathbf{D_1}\right) + \left({}^0\mathbf{D_2'} - {}^0\mathbf{D_2}\right) + \left({}^0\mathbf{D_3'} - {}^0\mathbf{D_3}\right) = 0 \tag{4.61}$$

The incidence matrix can be used to find a unique labeling set.

4.5.3 Consistent Updating of a Location Network

Consider taking a sequence of observations $\vec{z} = \{z_1, \cdots, z_k\}^T$ of the locations of centroids in the world model, found by updating from sensor feature measurements. We will assume that prior clustering of the observations has occurred so that for each arc with description vector D_i, there is at most one observation z_i which admits a Bayesian consensus with D_i. Initially, consider the case where each observation is independent and modeled by a pure Gaussian $f(z_i \mid D_i) \sim N(z_i, \Gamma_i)$. Further assume that the prior marginal distributions $\pi_i(D_i)$ are Gaussian as $N(D_i, \Lambda_i)$ so that the posterior distributions of the arc labels are given by

$$\pi_i(D_i \mid z_i) \sim N\left(\overline{D}_i, \overline{\Lambda}_i\right) \tag{4.62}$$

where

$$\overline{D}_i = \left(\Gamma^{-1} + \Lambda_i^{-1}\right)^{-1}\left(\Lambda_i^{-1}D_i + \Gamma_i^{-1}z_i\right) \tag{4.63}$$

and

$$\overline{\Lambda}_i = \left(\Gamma^{-1} + \Lambda_i^{-1}\right)^{-1} \tag{4.64}$$

Following Section 4.3, we will use a maximum posteriori estimator for D_i'. From Equation 4.15, the value of D_i' that maximizes Equation 4.62 also minimizes the quadratic

$$
\begin{aligned}
L_i(D_i') &= \tfrac{1}{2}\left(D_i' - \overline{D}_i\right)^T \overline{\Lambda}_i^{-1}\left(D_i' - \overline{D}_i\right) \\
&= \tfrac{1}{2}(D_i' - D_i)^T \Lambda_i^{-1}(D_i' - D_i) \\
&\quad + \tfrac{1}{2}(D_i' - z_i)^T \Gamma_i^{-1}(D_i' - z_i)
\end{aligned}
\tag{4.65}
$$

Note that $L_i(D_i')$ is location invariant, so that it has the same interpretation in *any* coordinate frame. This can be verified by transforming the estimate vectors and the variance matrix in the quadratic by the transform and the Jacobian respectively. Consequently we can consider minimization of 4.65 in any convenient coordinate frame. In the case where the location observations z_i are only partial (for example, where depth alone is being measured), the associated elements of Γ_i^{-1} are set to zero as is (arbitrarily) the unobserved elements of the z_i vector. The result of this is that a zero weight is attached to such elements of the observation in Equation 4.65 so that they are ignored.

We can now restate the problem of finding a consistent fusion of observation vectors z_i into a prior network of locations as finding the location labels $\hat{D}_i$ that minimize $L_i(\cdot)$, $i = 1, \cdots, m$, subject to $C(^0\hat{D}_i - {}^0D_i) = 0$ in a fixed but arbitrary coordinate frame. We can summarize these requirements into a common loss function. Define

$$^0\hat{D} = \left[{}^0\hat{D}_1^T, \cdots, {}^0\hat{D}_1^T\right]^T \tag{4.66}$$

and let

$$L(^0\hat{D}) = \sum_{i=1}^{m} L_i(^0\hat{D}_i) \tag{4.67}$$

As each L_i is a positive quadratic, minimizing each L_i individually with respect to the appropriate $^0\hat{D}_i$ is equivalent to minimizing $L(\cdot)$ with respect to the vector $^0\hat{D}$. Define $^0\Sigma_D$ ($6m \times 6m$) to be the diagonal matrix of arc location variance-covariance matrices $^0\Lambda_i$ ($i = 1, \cdots, m$) in this common coordinate frame, and define Γ_z ($6k \times 6k$) to be the (non-singular but not necessarily diagonal) joint variance-covariance matrix of the observation vectors $z = [z_1, \cdots, z_k]^T$ ($k \leq m$). Define N_z ($6k \times 6m$) as the observation incidence matrix that gives $(N_z{}^0D - z)$ a sensible interpretation in the appropriate coordinate frame. We can now restate the update problem as one of finding the vector $^0\hat{D}$ that minimizes:

$$\begin{aligned} L(^0\hat{D}) \;=\; & \tfrac{1}{2}\left(^0\hat{D} - {}^0D\right)^T \Sigma_D^{-1} \left(^0\hat{D} - {}^0D\right) + \\ & \tfrac{1}{2}\left(N_z{}^0\hat{D} - z\right)^T \Gamma_z^{-1} \left(N_z{}^0\hat{D} - z\right) \end{aligned} \tag{4.68}$$

subject to
$$G(^0\hat{D}) = C\left(^0\hat{D} - {^0D}\right) = 0 \qquad (4.69)$$
Differentiating Equations 4.68 and 4.69 and using Lagrangian multipliers as
$$\frac{\partial L}{\partial^0\hat{D}} = \frac{\partial G}{\partial^0\hat{D}}\Lambda \qquad (4.70)$$
we obtain
$$\Sigma_D^{-1}\left(^0\hat{D} - {^0D}\right) + \Gamma_z^{-1}\left(N_z{^0\hat{D}} - z\right) = C^T\Lambda \qquad (4.71)$$
Multiplying by the incidence matrix M, recalling that $MC^T = 0$ and simplifying gives:
$$M\left[\Sigma_D^{-1} + N_z^T\Gamma_z^{-1}N_z\right]\left(^0\hat{D} - {^0D}\right) = MN_z^T\Gamma_z^{-1}\left(z - N_z{^0\hat{D}}\right) \qquad (4.72)$$
This equation yields the required changes in description vectors $(^0\hat{D} - {^0D})$ in terms of the difference between observations and prior information $(z - N_z{^0\hat{D}})$, all referenced to a common coordinate frame. The resulting update in relations between objects following the fusion of the observations z, maintain external consistency and are internally Bayes-rational.

Equation 4.72 represents $6n$ equations in $6m$ $(^0\hat{D} - {^0D})$ unknowns, $m \geq n$. However, as before, the network structure ensures that the elements of the vector $(^0\hat{D} - {^0D})$ are not independent, indeed there are only $6(m-1)$ independent unknowns. To create an independent set we choose a new arbitrary but fixed coordinate frame and connect all n objects in the world model network to this frame with location vectors x_i, $i = 1, \cdots, n$. The vector $x = [x_1^T, \cdots, x_n^T]^T$ is related to the vector of description vectors by $^0D = M^T x$, and similarly $^0\hat{D} = M^T\hat{x}$, so
$$(^0\hat{D} - {^0D}) = M^T(\hat{x} - x) \qquad (4.73)$$
Then Equation 4.72 can be written as
$$M\left[\Sigma_D^{-1} + N_z^T\Gamma_z^{-1}N_z\right]M^T(\hat{x} - x) = MN_z^T\Gamma_z^{-1}\left(z - N_z{^0\hat{D}}\right) \qquad (4.74)$$

representing $6n$ equations in the $6n$ unknowns $(\hat{x}-x)$ By Result 2.2 of Chapter 2, the matrix

$$\mathbf{M}\left[\Sigma_D^{-1} + \mathbf{N}_z^T\Gamma_z^{-1}\mathbf{N}_z\right]\mathbf{M}^T \tag{4.75}$$

is singular (rank $6(n-1)$) and so one of the x_i is arbitrary and must be fixed. It is usually easiest to choose the coordinate frame in which x is defined as being coincident with an object which is part of the observation. For example, if a relation to the object O_j is observed, set $(\hat{x}_j - x_j) = 0$ and solve for the remaining x_i from Equation 4.72. The updated network joint variance covariance matrix is just given by Equation 4.75.

4.5.4   Computational Considerations

It is important to analyze the complexity of Equation 4.74. In the general case, the solution of the consistency problem has a complexity which is exponential in the number of objects in the world model. However, there are two methods by which computation can be reduced. Firstly the constraint equations only apply around loops of relations, and secondly the effect of an update diminishes with the number of relations through which it is propagated.

Consider the schematic diagram [Figure 4.3] of a general network description of part of a world model. Each node represents an object (polygon, patch, sensor, frame, etc), and each edge is a relation between objects. In this example, consider the network as representing two objects, related to each other, both connected to a coordinate frame, and both described by a set of geometric primitives (surface patchs for example). As there are no direct relations between primitives describing object 1 and the primitives describing object 2, there are no network loops between them. So updating any of object 1's primitives will *not* change any of the primitives describing object 2. Updating a relation will only affect those relations that form a loop with the observation. In Figure 4.3, nodes 1 and

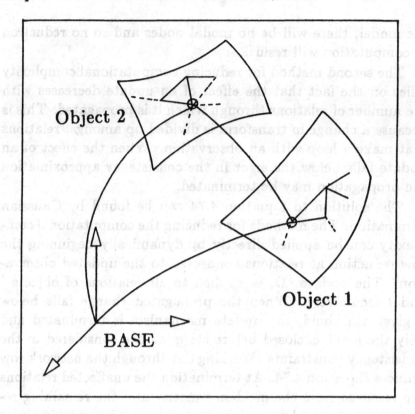

Figure 4.3: Schematic Diagram of a General Object Topology

2 form "modal nodes", though which changes are not propagated. These modal nodes divide up the incidence matrix in to disjoint sub-matrices, so that the effect of an update is restricted to the sub-incidence matrix in which it was first applied. There are two notable extremes of this property. Firstly in a *pure* hierarchy, there are *no* relational loops, so that each relation is its own sub-incidence matrix and thus there is no propagation of information or need to maintain consistency. Indeed the fact that there are no loops and thus no propagation, implies that a pure hierarchical representation does not make full use of the available information. Secondly in the unlikely case that all objects are related to all other objects in

the model, there will be no modal nodes and so no reduction in computation will result.

The second method for reducing computational complexity relies on the fact that the effect of an update decreases with the number of relations through which it is propagated. This is because a change in transform is divided up amongst relations that make a loop with an observation. When the effect of an update falls below the error in the consistency approximation the propagation may be terminated.

The solution to Equation 4.74 can be found by Gaussian elimination. The methods for reducing the computational complexity can be applied directly by dynamically beginning the row reduction at relations connected to the updated observation. The update ^jD_i is applied to all relations of objects j and i recursively. When the propagated change falls below a given threshold, the update mechanism is terminated and only the loops enclosed before stopping are considered in the consistency constraints. Working out through the network row reduces Equation 4.74. At termination the unaffected relations are removed from the incidence matrix and the remaining reduced rows form a lower triangular matrix. Then working back from the termination points, the appropriate values of D' may be calculated from the reduced matrix. As propagation generaly terminates in three applications, this represents a computationally simple implementation of the propagation algorithm.

4.6   Summary

In this Chapter, we have developed a method for comparing and combining sparse sensor observations, and for integrating this information into a prior network of geometric constraints. The integration mechanism relies on a probabilistic description of geometric observations made by different sensors in a multi-sensor systems. These observations are interpreted in terms of a known world model, then clustered to provide updates to estimated object locations. The resulting changes in object location are integrated into a network of constraints describing the relations between objects in the world. This integration mechanism provides a means of propagating new sensor observations through the robot systems world model, allowing consistency between prior constraints and observed information to be maintained.

Chapter 5

COORDINATION AND CONTROL

5.1 Introduction

Our development of multi-sensor systems has so far concentrated on the problem of integrating diverse geometric observations from different sensors to provide a robust estimate of the location of objects in a known but uncertain environment. To extend these techniques to the more complex problem of estimating the state of an unknown environment, we must make full use of the dynamic capabilities of sensor systems. These dynamic capabilities result from the exchange of information, by communication of action, between different sensors, each taking limited or partial observations of the environment. This information exchange provides a basis through which individual sensors can *cooperate* with each other, resolve *conflicts* or disagreements, or *compliment* each others view of the environment. The dynamic communication of information in this way is an essential element of the power provided by multi-sensor systems. To fully understand and utilize these capabilities, we must be able to coordinate the exchange of information between different observers, and control the acquisition of in-

141

formation by disparate sensory systems. This requires that we model the relations between information provided by different sensors, and that we describe how this information is obtained. In Chapter 3, we introduced the idea of an information structure as a model for the dynamic acquisition and exchange of different sensor opinions. We will now use this model to develop a means to coordinate and control sensory systems.

We will describe a multi-sensor system as a *team* of decision makers. Each sensor of this team will be considered as an individual decision maker; taking observations, making local decisions and implementing it's own actions. Together, the sensors must coordinate their activities, guide each other to view areas of interest, and ultimately come to some team-consensus view of the environment. The characteristics of a team structure capture many of the desirable properties of a multi-sensor system: Team members can make local decisions based on a team goal, providing a means of delegating observation tasks and actions. Team members can help each other by exchanging information unattainable to individual members, thus providing the team with a more complete description of events. Team members can disagree with each other as to what is being observed; resolving differences of opinion can supply a mechanism for validating each others operation. The most important observation about a team is that the coordinated activities of the members is more effective than the sum of their individual actions.

The description of a multi-sensor system as a team of decision makers contains a number of important elements;

- How individual sensors make local decisions based on their own observations.

- How the capabilities and information provided by each sensor are described to other members of the team.

- How information obtained by one sensor is described and

communicated to other sensors.

- How local decisions made by individual sensors can be integrated to provide a team decision and action

We will develop a description of a sensor as a Bayesian observer of geometric states in the environment. The capabilities of each sensor will be represented by an information structure, describing the observation of geometric features with respect to the state of the sensor and the availability of prior information. In this framework, local decisions are made with respect to an individual utility function. An individual team members decision is chosen to maximize it's local utility. If a sensor is considered as a Bayesian observer, this reduces to finding an estimate of the observed geometric feature which maximizes likelihood.

Each individual sensor or team member communicates information in the common language of uncertain geometry. This allows different sensors to exchange information and utilize diverse opinions from other, different, sources. Each sensor must also communicate a preference order on it's decisions to provide a means of resolving any differences of opinion amongst sensors. In a team of Bayesians, this preference order is described by a likelihood function. The description of preferences in terms of a probability distribution provides a dimensionless means of comparing estimates of disparate geometric features, in a common framework. The communication of observations and preferences forms the basis for coordinating the acquisition of information, and provides a mechanism for the generation and verification of team hypotheses.

The observations and decisions made by different sensor comprising a multi-sensor team must be integrated to provide a team decision and action. This requires that we place a preference ordering on team decisions and describes how individual members observations and actions affect the team consensus. In a multi-Bayesian system of observers, we must develop a mechanism for describing the combination of information, the differences of opinion amongst individuals, and the formation of coalitions, guiding and modifying the team estimates. This mechanism characterizes the team decision problem.

In this Chapter, we will develop a description of a multi-sensor system as a team of decision makers. Section 5.2 introduces the key elements of team decision making. The description of individual decision makers is developed in terms of an information structure and a preference order on decisions. The problem of assigning group rationality is described and the multi-Bayesian system of observers is introduced. This description is then used to provide a mechanism for the coordination and pooling of diverse sensor opinions. Section 5.3 develops the idea of a multi-sensor team in the framework of a multi-Bayesian system. The process of decision making in unknown environments is described, and techniques are developed to generate and verify team decisions based on local, partial sensor observations. The coordination of this decision making process is developed with respect to a network of sensors, communicating information, generating decisions and constraining each others view of the environment. This description of information is also used to consider the problem of controlling and guiding sensors. This allows the development of sensing strategies in the context of individual control and team decision making.

5.2   The Team Decision Problem

We will consider a multi-sensor team structure consisting of n individual sensory systems. Each team member is represented by an information structure $\eta_i(\mathbf{x}_i, \mathbf{p}, \overline{\delta}_i)$, that describes the dependence of each sensors observations on state, prior information, and the decisions made by other sensors in the team. The n-tuple $\vec{\eta} = \{\eta_1, \cdots, \eta_n\}$ is called the team information structure.

The observations z_i made by different sensors can be described, in terms of uncertain geometry, as a probability distribution function. Each sensor must make a decision δ_i, based on these observations, to estimate some geometric feature in the environment $\delta_i(z_i) \in P_i$. The n-tuple $\vec{\delta} = \{\delta_1, \cdots, \delta_n\}$ is called the team the team decision function.

The decisions made by each sensor must be evaluated with respect to some measure of preference. This preference ordering, or utility function $u_i(\cdot, \delta_i) \in \Re$, allows differences of opinion between sensors to arise, conflicts to be resolved and local decisions combined into global results. Consider for example, two sensors taking observations z_1 and z_2 of two physically disparate geometric features p_1 and p_2 respectively. Suppose that the observed features are related in some way so that the estimate $\delta_1(z_1)$ is constrained by the estimate $\delta_2(z_2)$. In general, these two estimates cannot be compared directly and so the constraints between them cannot be made explicit. However, we can compare the contribution or utility they each provide to some joint consensus. Thus by first choosing some consensus estimate, we can compare diverse opinions by evaluating $u_i(\cdot, \delta_i(z_i))$. The local decision is to maximize u_i, the group decision is to encorporate geometric constraint between sensors in this maximization. This comparison is dimensionless, allowing all geometric features to be considered in a common framework. Our interset now centers on when a constrained estimate satisfying all sensors exists, and what consensus value it will take. This is the team decision problem.

5.2.1 The Structure of a Team

A sensor or team member will be considered *rational* if for each observation z_i of some prior feature $p_i \in P_i$ it makes the estimate $\delta_i(z_i) \in P_i$ which maximizes it's individual utility $u_i(p_i, \delta_i(z)_i) \in \Re$. In this sense, utility is just a metric for constructing a complete lattice of decisions; allowing any two decisions to be compared in a common framework.

Individual rationality alone imposes insufficient structure on the organization of team members: If each sensor is allowed to make it's own decisions regardless of other sensor opinions, then a group consensus will never be reached. We must provide a structure which allows for sensor opinions to be constrained or biased towards a common, team objective. To do this, we will define a *team* utility function defined on the state of the environment $\mathbf{p}$ and the team decision function $\vec{\delta}$:

$$U = U(\mathbf{p}, \delta_1, \delta_2, \cdots, \delta_n) \tag{5.1}$$

This is a real-valued function which for every team decision $\vec{\delta}$ assigns a team utility, admitting a preference (in some sense) on group decisions. The fundamental problem in team decision theory is based on finding this preference ordering on $\vec{\delta}$. This problem, the formulation of "Group Rationality Axioms", can be quite complex and is still an open research issue (see for example [Harsanyi 76, Bachrach 75]).

The role of U is very important in characterizing the structure of a sensor team. The interpretation of team action due to Ho and Chu [Ho 72], or Marshak and Radnor [Marshak 72], is to require each team member to make the decision δ_i that maximizes the team utility U, regardless of personal preference. We will call this an *altruistic* team structure. In the scenario of a multi-sensor system, it may not always be advisable for single sensors to agree with a team decision, particularly if the sensor has for some reason made a mistake and should *not* be contributing to a consensus decision. To account for this, an alternative formulation is to allow individual team members to have a personal utility as well as an interest in the team. For example, a team member may agree to cooperate and be subject to the utility U, or to disagree with the other team members and be subject to a personal utility u_i. In this case a rational team member will agree to cooperate only if it will gain by doing so, when the team utility exceeds its personal utility. We shall call this an *antagonistic* team.

We can considerably simplify the team decision problem by restricting the type of group preference orderings that can be considered "rational". Consider a multi-sensor team taking observations of a number of different features in the environment. Each sensor has a preference ordering on the decisions that it can make; if it estimates the location of a feature, then it would prefer to have this estimate verified rather than disproved or relocated by the group decision. Clearly it must allow its decisions to be changed by at least a small amount, so that it can be brought in to line with a consistent interpretation of all sensors observations. However, it should not allow its opinion to be changed so much that the group opinion now bears no resemblance to the sensors original observations. This dichotomy between consensus and disagreement is fundamental to the multi-sensor decision problem.

If the sensor system suggests an explanation for all the different sensor observations, then each individual sensor must interpret this decision in terms of an explanation for its own measurements. If the sensor cannot reconcile its views with those of the team, then it must disagree and not support the team decision. If however it can support the team interpretation of sensor observations, the sensor should reflect this in its individual preference ordering. In this situation, the sensor system will try to find decisions that explain as many different sensor observations as possible, suggesting consistent interpretations for the different opinions, but allowing sensors to disagree. If sensors were not allowed to disagree, then they may be required to support hypothesis that are actually incorrect. The rôle of the sensors and sensor system in this case is to dynamically find a consistent, consensus estimate of the state of the environment by alternately suggesting and trying to verify possible environment hypothesis. This process of dynamic interaction, agreement and disagreement is called the *bargaining* problem [Nash 50]

5.2.2   The Multi-Bayesian Team

We are primarily interested in teams of observers; sensors making observations of the state of the environment. In this case individual team members can be considered as Bayesian estimators. The team decision is to come to a consensus view of the observed state of nature. The static team of estimators is often called a Multi-Bayesian system [Weerhandi 81,83]. These systems have many of the same characteristics as more general team decision problems, but considerably simplify the bargaining problem by interpreting group rationality as seeking only the combined group maximum likelihood estimates of the observations. A multi-Bayesian team works by considering the likelihood function $f_i(\cdot \mid \mathbf{p})$ of each team member as the (normalized) utility of individual observers. Then the team utility is considered to be the joint posterior distribution function $F(\mathbf{p} \mid z_1, \cdots, z_n)$ after each sensor i has made the observation z_i. The advantage of considering the team problem in this framework is that both individual and team utilities are normalized so that comparisons can be performed easily, supplying a simple and transparent interpretation to the group rationality problem.

Before developing the general multi-Bayesian system, it is helpful to study the simpler case of two scaler homogeneous observers, in which many of the important problems in group decision making can be analyzed: Consider two team members, each observing the same scalar variable $p \in P$ (feature in the environment), with observation density $f_i(\cdot \mid p)$, $i = 1, 2$. Suppose each observer takes a single observation z_i, considered independent and derived from a Gaussian distribution with mean $\hat{p}$ and variance σ_i. The goal of this team is to come to some consensus estimate of state $\overline{p} \in P$, based on the two observations z_1 and z_2. In the multi-Bayesian system, each team members individual utility function is given by the posterior likelihood;

$$u_i(\bar{p} = \delta_i(z_i), p) = f(p \mid z_i) \sim N(\hat{p}, \sigma_i) \qquad i = 1, 2.$$

The team utility function is given by the joint posterior likelihood;

$$U(\bar{p}(z_1, z_2), p) = F(p \mid z_1, z_2)$$
$$= f_1(p \mid z_1) f_2(p \mid z_2)$$

A team member will be considered *individually* rational if it chooses the estimate $\bar{p} \in P$ which maximizes it's local posterior density;

$$\bar{p} = \arg \max_{p \in P} f_i(p \mid z_i) \qquad i = 1, 2. \tag{5.2}$$

The team itself will be considered group-rational if together the team members choose the estimate $\bar{p} \in P$ which maximizes the joint posterior density;

$$\bar{p} = \arg \max_{p \in P} F(p \mid z_1, z_2)$$
$$\tag{5.3}$$
$$= \arg \max_{p \in P} f_1(p \mid z_1) f_2(p \mid z_2)$$

In the first case, described by Equation 5.2, the estimate will just be the unique mode of the (Gaussian) posterior likelihood function; exactly as for an unbiased Bayes estimator. However, in the second case, described by Equation 5.3, two possible results can be obtained;

1. $F(p \mid z_1, z_2)$ has a unique mode equal to the estimate $\bar{p}$ in Equation 5.3.

2. $F(p \mid z_1, z_2)$ is bimodal and no unique group-rational consensus estimate exists.

If $F(p \mid z_1, z_2)$ has a unique mode, it will satisfy;

$$\max_{p \in P} F(p \mid z_1, z_2) \geq \max_{p \in P} f_i(p \mid z_i); \qquad i = 1, 2. \qquad (5.4)$$

Conversely, if $F(p \mid z_1, z_2)$ is bimodal then

$$\max_{p \in P} f_i(p \mid z_i) > \max_{p \in P} F(p \mid z_1, z_2); \qquad i = 1, 2. \qquad (5.5)$$

We will consider our two-member team to be antagonistic in the following sense: Each team member can either agree with a team estimate $\bar{p}_t$ and be subject to the team utility

$$F(\mathbf{p} = \bar{p}_t \mid z_i, \cdot),$$

or choose it's own estimate $\bar{p}_i$ and be subject to an individual utility

$$f_i(\mathbf{p} = \bar{p}_t \mid z_i).$$

A rational team member will maximizes utility by choosing to either agree or disagree with the team consensus: If a team members observation satisfies Equation 5.5, then it will not cooperate with the team estimate. Thus an antagonistic team member will be rational if it makes the decision;

$$\begin{aligned}
\bar{p} &= \delta_i(z_i) \\
&= \arg\max_{p \in P}\{f_i(p \mid z_i), F(p \mid z_1, z_2)\}; \qquad i = 1, 2.
\end{aligned}$$

Whether or not the individual team members will arrive at a consensus team estimate will depend on some measure of how much they disagree $|z_1 - z_2|$. If z_1 and z_2 are "close enough" then the posterior density $F(p \mid z_1, z_2)$ will be unimodal and satisfy Equation 5.4, with the consensus estimate given by equation 5.3 (Figure 5.1). As $|z_1 - z_2|$ increases, $F(p \mid z_1, z_2)$ becomes flatter and eventually bimodal (Figure 5.2). At this point, the joint density will satisfy Equation 5.5, no consensus team decision will be reached, and the members individual estimates will satisfy Equation 5.2.

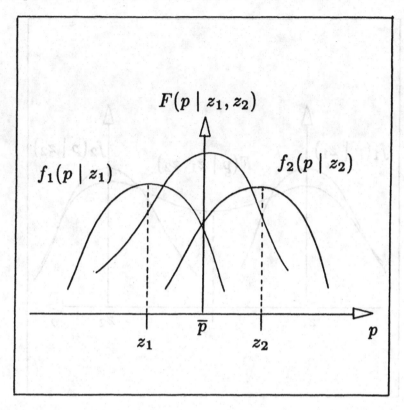

Figure 5.1: Two Bayesian observers with joint posterior likelihood indicating agreement.

We are now interested in finding the point at which two antagonistic Bayesian team members will disagree with each other. From Equations 5.4 and 5.5, the point of disagreement will occur when the joint posterior density $F(p \mid z_1, z_2)$ ceases to be convex on the interval $[z_1, z_2]$. To test for this convexity, we need only ensure that the second derivative of the function

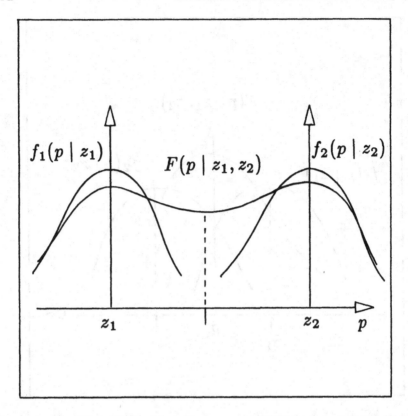

Figure 5.2: Two Bayesian observers with joint posterior likelihood indicating disagreement.

$F(p \mid \cdot)$ is positive. Differentiating, we obtain:

$$\frac{\partial^2 F}{\partial p^2} = \frac{1}{f_1}\frac{d^2 f_1}{dp^2} + \frac{1}{f_2}\frac{d^2 f_2}{dp^2} + \frac{2}{f_1 f_2}\frac{df_1}{dp}\frac{df_2}{dp}$$

$$= \left(\sigma_1^{-2} + \sigma_2^{-2}\right) - \left[\sigma_1^{-2}(p - z_1) + \sigma_2^{-2}(p - z_2)\right]^2$$

For this to be positive and hence $F(p \mid \cdot)$ to be convex, we are required to find a consensus p which satisfies:

$$\left[\sigma_1^{-2}(p - z_1) + \sigma_2^{-2}(p - z_2)\right]^2 \left[\sigma_1^{-2} + \sigma_2^{-2}\right]^{-1} \leq 1 \qquad (5.6)$$

Notice that Equation 5.6 is no more than a normalized weighted sum, a scalar equivalent to the Kalman gain matrix. It follows therefore, that if a p can be found which satisfies Equation 5.6, then the consensus $\overline{p}$ which maximizes F will be given by:

$$\overline{p} = \frac{\left(\sigma_1^{-2}z_1 + \sigma_2^{-2}z_2\right)}{\left(\sigma_1^{-2} + \sigma_2^{-2}\right)} \tag{5.7}$$

An equivalent test for the existence of a consensus is to consider the convexity of the *set* ;

$$\mathbf{f}(p) = [f_1(p \mid z_1), f_2(p \mid z_2)]^T \in \Re^2.$$

Substituting Equation 5.7 into Equation 5.6, we can write;

$$(z_1 - z_2)(\sigma_1^2 + \sigma_2^2)^{-1}(z_1 - z_2) = D_{12}(z_1, z_2)$$

where $D_{12} \leq 1$ in Equation 5.6. Figure 5.3 shows a plot of the set $\mathbf{f}(p)$ for two Bayesians. The disagreement measure $D_{12} = D_{12}(z_1, z_2)$ is called the Mahalanobis distance. Figure 5.3 shows that for $D_{12} \leq 1$, the set $\mathbf{f}(p)$ is convex and a consensus, given by Equation 5.7 exists. As the difference $|z_1 - z_2|$ increases, so D_{12} becomes larger and eventually the set $\mathbf{f}(p)$ becomes concave, indicating disagreement. The consensus values of p describes an envelope of extrema of $\mathbf{f}(p)$. This provides an important visualization of agreement or disagreement between a set of Bayesians.

We will now generalize this two–Bayesian system to a team of n Bayesians all taking different observations of geometric features in the environment. Consider n sensors each taking observations $\mathbf{z}_i \in P_i$ of different geometric features $\mathbf{p}_i \in P_i$; $P_i \neq P_j$. To compare geometrically disparate observations, we must hypothesize the existence of some common geometric object $\mathbf{p} \in P$ to which these observations are related. For example, this object could be the location of a centroid, or the description of a common surface. The observations made by a

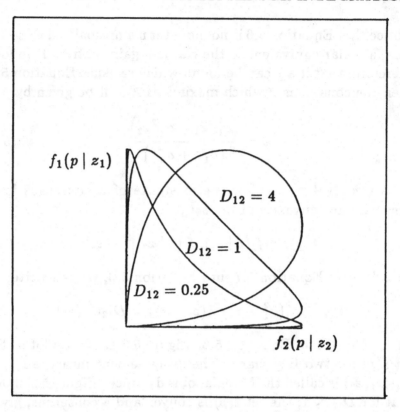

Figure 5.3: The space of preferences for a two-Bayesian team

sensor can be related to this common object through the decision function $\delta_i(z_i) \in P^1$. It is important to note that although $\delta_i^{-1}(p) \in P_i$ is usually well-defined, $\delta_i(z_i) \in P$ is often only a partial mapping. For example given the centroid of a known polyhedra, we can uniquely determine the location of all its edges and surfaces. Conversely given an observed edge of this polyhedra, it is not possible to uniquely determine the objects centroid without recourse to other constraining information. However, although the decision $\delta_i(z_i)$ contains only partial information about p, we can consider the observation in terms of

---

[1] We have included the transformation function h_i in the decision function δ_i to simplify notation.

uncertain geometry and thus manipulate or transform it using the techniques described in Chapter 2.

Following our development of the two-Bayesian team, we will consider an individual sensors preference ordering on consensus descriptions of a common geometric object $\mathbf{p}$ as the posterior density $f_i(\mathbf{p} \mid \delta_i(\mathbf{z}_i))^2$. This can be interpreted in two ways: If the parameter $\mathbf{p}$ is fixed, then f_i is considered as the support given by the observation $\mathbf{z}_i$ to the hypothesis parameterized by $\mathbf{p}$. Alternatively, if $\delta_i(\mathbf{z}_i)$ is considered fixed, then f_i describes the preference order that a sensor puts on possible values of $\mathbf{p}$. The first case corresponds to a centralist philosophy in which some sensor manager suggests hypotheses and seeks support from sensor observations. This approach would be appropriate for use in a known environment, model-based system. The second case corresponds to a decentralized decision making procedure in which a sensor suggests values for $\mathbf{p}$, the combination of which provide a consensus. This approach is more appropriate in situations where the environment is unknown.

As before, we will define the preference order on team decisions to be the joint posterior density function $F(\mathbf{p} \mid \cdot)$. If the sensors are considered to make independent observations, then this can be written as;

$$F(\mathbf{p} \mid \delta_1(\mathbf{z}_1), \cdots, \delta_n(\mathbf{z}_n)) = \prod_{i=1}^{n} f_i(\mathbf{p} \mid \delta_i(\mathbf{z}_i)) \qquad (5.8)$$

Following similar arguments to those of the two-Bayesian system, the observations $\mathbf{z}_i$ will only provide a consensus estimate for $\mathbf{p}$ if $F(\mathbf{p} \mid \cdot)$ is convex in all it's arguments. To determine convexity, we need only ensure that the Hessian of F is positive

---

[2]Note that when δ_i provides only partial information, f_i will be improper.

semi-definite. Differentiation provides;

$$\frac{\partial^2 F}{\partial \mathbf{p}^2} = \sum_{i=1}^{n} \left[\frac{1}{f_i} \frac{d^2 f_i}{d\mathbf{p}^2} \right] + \sum_{i=1}^{n} \sum_{\substack{j=1 \\ j \neq i}}^{n} \left[\frac{1}{f_i f_j} \left(\frac{df_i}{d\mathbf{p}} \right) \left(\frac{df_j}{d\mathbf{p}} \right)^T \right]$$

If each f_i is considered Gaussian with variance $\mathbf{\Lambda}_i$, this results in:

$$\frac{\partial^2 F}{\partial \mathbf{p}^2} = \sum_{i=1}^{n} \mathbf{\Lambda}_i^{-1} \left[\mathbf{I} - (\mathbf{p} - \delta_i(\mathbf{z}_i)) (\mathbf{p} - \delta_i(\mathbf{z}_i))^T \mathbf{\Lambda}_i^{-1} \right]$$

$$+ 2 \sum_{i=1}^{n} \sum_{j=i+1}^{n} \left[\mathbf{\Lambda}_i^{-1} (\mathbf{p} - \delta_i(\mathbf{z}_i)) (\mathbf{p} - \delta_j(\mathbf{z}_j))^T \mathbf{\Lambda}_j^{-1} \right]$$

$$= \left[\sum_{i=1}^{n} \mathbf{\Lambda}_i^{-1} \right]$$

$$- \left[\sum_{i-1}^{n} \mathbf{\Lambda}_i^{-1} (\mathbf{p} - \delta_i(\mathbf{z}_i)) \right] \left[\sum_{i-1}^{n} \mathbf{\Lambda}_i^{-1} (\mathbf{p} - \delta_i(\mathbf{z}_i)) \right]^T$$

To find what values of the parameter $\mathbf{p}$ make this positive semi-definite (p.s.d.), observe that this matrix is in the form $(\mathbf{\Gamma} - \mathbf{xx}^T)$. As $\mathbf{\Gamma}$ is p.s.d., it suffices to show that $(\mathbf{I} - \mathbf{\Gamma}^{-1}\mathbf{xx}^T)$ is also p.s.d. This is equivalent to demonstrating that for arbitrary vectors $\mathbf{y}$:

$$\mathbf{y}^T \left(\mathbf{I} - \mathbf{\Gamma}^{-1}\mathbf{xx}^T \right) \mathbf{y} \geq 0$$

In particular, let $\mathbf{y} = \mathbf{x}$, then

$$\mathbf{x}^T \left(\mathbf{I} - \mathbf{\Gamma}^{-1}\mathbf{xx}^T \right) \mathbf{x} = \left(1 - \mathbf{x}^T \mathbf{\Gamma}^{-1}\mathbf{x} \right) (\mathbf{x}^T\mathbf{x}) \geq 0$$

or

$$\mathbf{x}^T\mathbf{\Gamma}^{-1}\mathbf{x} \leq 1$$

Therefore, finding a consensus value of **p** which makes the Hessian of F p.s.d. is equivalent to finding a **p** which satisfies

$$\left[\sum_{i=1}^{n} \mathbf{\Lambda}_i^{-1}\left(\mathbf{p} - \delta_i(\mathbf{z}_i)\right)\right]^T \left[\sum_{i=1}^{n} \mathbf{\Lambda}_i^{-1}\right]^{-1} \left[\sum_{i=1}^{n} \mathbf{\Lambda}_i^{-1}\left(\mathbf{p} - \delta_i(\mathbf{z}_i)\right)\right] \leq 1$$

(5.9)

If it exists, the consensus parameter estimate $\hat{\mathbf{p}}$ that maximizes Equation 5.8 and minimizes the left side of Equation 5.9, is given by the modified Kalman minimum-variance estimate;

$$\hat{\mathbf{p}} = \left[\sum_{i=1}^{n} \mathbf{\Lambda}_i^{-1}\right]^{-1} \left[\sum_{i=1}^{n} \mathbf{\Lambda}_i^{-1}\delta_i(\mathbf{z}_i)\right]$$

(5.10)

Following our development of the two-Bayesian team, it is useful to interpret Equation 5.9 in terms of the set of posterior density functions;

$$\mathbf{f}(\mathbf{p}) = \left[f_1(\mathbf{p} \mid \delta_1(\mathbf{z}_1)), \cdots, f_2(\mathbf{p} \mid \delta_1(\mathbf{z}_n))\right]^T$$

We will call this set $\mathbf{f}(\mathbf{p}) \subseteq \Re^n$ the opinion space of the n-sensor team. This is an n-dimensional space with basis axis corresponding to individual sensor preferences f_i[3]. The set $\mathbf{f}(\mathbf{p})$ describes an $n-1$ dimensional surface in this space, parameterized by **p** and enclosing a volume $v(\mathbf{z}_1, \cdots, \mathbf{z}_n) \in \Re^n$. A consensus of *all* sensors requires that this volume be convex along each axis. The consensus value lies on the surface enclosing this volume. If $v \in \Re^n$ is concave along the jth axis then the jth sensor observation will be unable to agree with other sensor opinions; it's individual maximum likelihood estimate is preferred to the team decision. Thus if the volume v is concave in one or more directions f_i, then coalitions supporting different decisions will be formed. This is quite a natural thing to occur; individual sensor systems may indeed be viewing different features and so should not agree with a single consensus decision. Further, a partitioning of the f_i in to coalitions should tell

---

[3]Figure 5.3 is a two-sensor example of this space.

the sensor system that its suggested consensus is only partially correct, and should be altered to bring the different coalitions into a single, convex opinion.

The antagonistic team structure allows members to disagree if, for some reason, they have made a mistake or cannot reconcile their views with those of other team members. If the sensor system rejects the opinion of an individual team member, this can be used as an incentive for the sensor to review it's observation process, entering a dialogue aimed at resolving differences of opinion. An altruistic team structure would be unable to disagree and enter this dynamic exchange of information.

This development of the n-sensor Bayesian team decision problem demonstrates three important principles;

1. Sensors, making different observations, generate partial hypotheses on group decisions which can be used to confirm or verify a team consensus.

2. The pooling of disparate opinions provides a mechanism for obtaining information unavailable to single decision makers.

3. The ability of sensors to disagree with each other in arriving at a consensus view of the environment is a natural and powerful means of constraining the search for an interpretation of observations while providing a mechanism to resolve differences in opinion.

The Bayesian-team structure will form the basis for our analysis of the coordination and control of multi-sensory systems.

5.2.3   Opinion Pools

The process of bringing together and combining information is called an opinion pool. In a multi-Bayesian system, this pooling is accomplished by using likelihood as a measure of both individual and team preferences. This is not the only way of combining different opinions, indeed likelihood may not always be a good measure of a team members preference. However, in this section, we shall show that the combination of preferences used so far, is the only reasonable choice for an opinion pool.

To make group decisions in the general team framework, we must extend the natural axioms of individual utility, that place a preference order on a single members decisions, to a measure of preference on group decisions. A set of group rationality axioms were first introduced by Nash [Nash 50]. There has been considerable disagreement about these axioms [Winkler 81], and a number of other definitions have been suggested (notably, Harsanyi [Harsanyi 77], devotes most of a book to this problem). To develop this team preference ordering, or team utility, individual decision makers must be able to compare their utilities using a common metric. Unfortunately, this comparison is not always possible; when two utilities do not have a comparable scale, when a combination of two or more utility functions is not itself a partial ordering, or when one utility presupposes some subjective knowledge of another members decisions.

The group decision problem can be considerably simplified by assuming the existence of a "disinterested" manager (Harsanyi's "ethical supervisor"), which can provide an unbiased mechanism for the comparison of individual utilities. The manager has complete knowledge of all members utility and can freely trade-off opinions, loyalty and incentives. The key principle in this form of group decision making is the idea of Pareto-optimal decision rules:

Definition: The group decision $\vec{\delta}^*$ is Pareto-optimal if every other rule $\vec{\delta} \in \mathcal{D}$ decreases *at least* one team members utility.

If the team utility $U(\mathbf{p}, \delta_1(\mathbf{z}_1), \cdots, \delta_n(\mathbf{z}_n))$ is convex, then it can be shown [Ho 72] that the Pareto-optimal team decision is also person-by-person optimal with respect to U. We can identify U with the opinion space $\mathbf{u} \subseteq \Re^n$ described in the previous section. We maintain that the goal of our sensor system is to provide a decision or interpretation of sensor observations that makes the opinion space convex in as many directions as possible. This allows disparate sensors to offer opinions in a common dimensionless framework and to disagree or guide the sensor system to a correct interpretation. The way in which individual preferences can guide the sensor system is dependent on the team utility. That is, the team utility is a function of the individual sensor decisions, therefore the team decision, which is based on this utility, is also dependent on individual decisions and preferences. Consider the team utility $U(\mathbf{p}, \delta_1 \cdots, \delta_n)$ as a function of the individuals opinion space $\mathbf{u} = [u_1(\delta_1), \cdots, u_n(\delta_n)]^T$, so that:

$$U = U(u_1, \cdots, u_n) = U(\mathbf{u})$$

Each individual utility can guide the team utility by expressing its preference to a team decision. In this way, the individual team members can guide the team decisions toward a group consensus that includes their own opinions. Conversely from the systems point of view, the role of the team utility U is to provide incentives for getting the individual team members to agree with each other. If the team is to exhibit these characteristics, we must restrict the possible forms the team utility can take to the following:

- Unanimity: $\frac{\partial U}{\partial u_i} > 0$, $\forall i$

- No dictator: If $\forall i: u_i \neq 0$, there is no u_j such that $U = u_j$.

- Indifference: If $\exists \delta_1, \delta_2$, $\forall i$ such that $u_i(\delta_1, \cdot) = u_i(\delta_2, \cdot)$, then $U(\cdot, \delta_1) = U(\cdot, \delta_2)$

The unanimity condition ensures that the team opinion moves in the same direction as individual preferences. The dictator condition requires that no one sensor can completely dominate the group decision at the expense of all other sensors. The indifference condition just states that if a sensor has no opinion to offer, then the team decision should be indifferent to that sensors preference. If the team utility function U can now be identified as an "opinion pool". The application of the three conditions restrict the form taken by the opinion pool to be a linear combination of either the generalized Nash product:

$$U(\mathbf{p}, \delta_1, \cdots, \delta_n) = c \prod_{i=1}^{n} u_i^{\alpha_i}(\mathbf{p}, \delta_i); \quad \alpha_i \geq 0$$

or the logarithmic or linear opinion pool:

$$U(\mathbf{p}, \delta_1, \cdots, \delta_n) = \sum_{i=1}^{n} \lambda_i u_i(\mathbf{p}, \delta_i); \quad \lambda_i \geq 0, \quad \sum_{i=1}^{n} \lambda_i = 1$$

The multi-Bayesian formulation of the team decision problem results in a specific case of the generalized Nash product with $\alpha_i = 1$ and individual utility being the posterior density.

5.3 Multi-Sensor Teams

We will now apply our development of multi-Bayesian teams to the problem of dynamic coordination and control of a multi-sensor system. We will consider each sensor or sensory cue as a member of an antagonistic Bayesian team. Each team member takes uncertain observations of disparate geometric features in the environment. These observations must be combined, regardless of the existence of prior information, into a consensus description of the environment. The process of observation and decision is inherently dynamic; what information should be provided by the sensor, where should the sensor locate itself to acquire appropriate observations, and how should the resulting information be used? We will maintain that the description of a multi-sensor system as a team of Bayesian decision makers provides a powerful means to describe this dynamic process, and answer these questions.

Consider the multi-sensor system comprising a set of sensors $\{S_1, \cdots, S_n\}$, each taking observations

$$\{z_i\} = \{z_{i,1}, \cdots, z_{i,m}\}$$

of different geometric features in the environment. We will assume that each sensor S_i observes only one type of geometric feature; $g(x, p_i) = 0$, $p_i \in P_i$. The observations $\{z_i\}$ made by the sensor are used to estimate instances of this type of feature: Each observation z_i provides an estimate $\delta_i(z_i) \in P_i$, a partial hypothesis of a feature instance $p_i \in P_i$. Thus the set of observations $\{z_i\}$ made by each sensor provide a set of feature hypotheses $\delta: \{z_i\} \mapsto \{p_i\}$. These hypotheses are uncertain, usually partial and often wrong.

We must make use of both prior information and the hypotheses generated by other sensors $\{p_j\}$ to improve our knowledge of the set $\{p_i\}$. To do this, we must provide some basis for comparing the information obtained by each sensing source. This not only requires a comparison metric, but also knowledge of how these different features are related. In the case of a known environment, the existence of prior information ensures that possible relations between observed features are known. In the case of an unknown environment, we must postulate the existence of some elemental global geometric description of the environment, and use this to evaluate the possible relationships between observed features. For example, we could assume that the environment is composed entierly of piece-wise quadratics (!), then the possible relations between observed geometric features will be determined by their contributions to specific quadratic descriptions.

The hypotheses $\{p_i\}$ generated by each sensor from it's observations can be interpreted in terms of partial hypotheses $\{p\}$ of some underlying global geometric environment description $g(x, p) = 0$, $p \in P$. The fact that individual sensor hypotheses

p_i can only ever supply partial estimates of the global geometry p, is a primary motivation for the use of many sources of sensory information. The partial hypotheses on global geometry provided by the sensors provides a means of comparing and combing different sensor views.

It would be naivé to allow sensors to generate and verify information all together (if only on computational grounds). Sensors should also be allowed to provide information directly to other sensors, to locally guide and resolve hypotheses verification. This exchange of partial, uncertain information between different sensor cues is an intrinsicly dynamic process. The information provided by one sensor is used by other sensors to localize their own observations, confirm or disprove hypothesis and in turn to supply information back to the original sensor for further localization. This dynamic interaction or dialogue serves to successively localize opinions, resolve ambiguities in single cue systems, and to provide robust convergence to a system consensus.

The information structure model of sensor behavior, described in Chapter 3, and our development of Bayesian teams in the previous section, provide a means of analyzing this process of dynamic information exchange.

5.3.1 Known and Unknown Environments

For the purpose of sensor integration, the most important difference between known and unknown environments is the existence of prior information with which to determine the relation between different features observed by the sensors. In a known environment, prior information about global geometry $\{p\}$ can be used to generate initial estimates of the features that a sensor can expect to observe. In this case, the interaction between different sensors serves to disambiguate local observations and provide confirmation or redundant information to isolate and resolve spurious or incomplete opinions. In an unknown environment, we must build descriptions of the global geometry directly from sensor observations, without reference to an existing prior world model. In this case, the interaction between different sensors serves three main purposes; reducing the search space of one sensor with another sensors observations, using a second sensor to confirm or disprove hypotheses generated by the first sensor, and using a second sensors information to prune the possible interpretations of a single sensors observations.

We can considerably simplify the process of interaction between sensors in an unknown environment by thinking of each observation as suggesting some partial interpretation of an underlying environment geometry. Each sensor can then be considered as generating partial elements of an *unknown* prior world model. These hypotheses can then be verified or rejected by comparison with the partial world models generated by other sensors observations. Subsequent observations can then be used to refine the estimated prior world model, in exactly the same manner as for a known environment.

The importance of identifying the similarity between prior information in a known environment and the hypotheses on underlying geometry generated by different sensors in an unknown environment, is that the analysis of both problems can be developed in exactly the same way. In each case, we assume the existence of a global geometry $\{p\}$, $g(x, p) = 0$. The features $\{p_i\}$, estimated by each sensor are then interpreted in terms of this common geometry, and the resulting partial estimates of $\{p\}$ used to resolve differences of opinion. This provides a means of relating, comparing and combining different sensor opinions in both known and unknown environments.

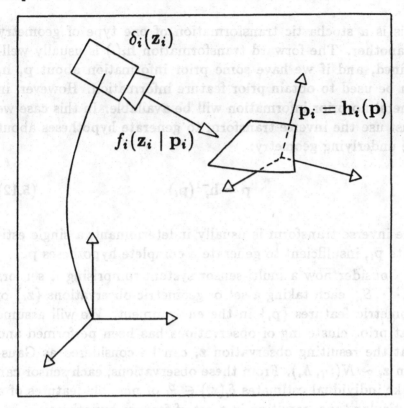

Figure 5.4: A single sensor observing a specific feature of a global geometric object

5.3.2   Hypothesis Generation and Verification

Consider a single sensor taking observations $\{z_i\}$ of a particular type of feature in the environment $g_i(x, p_i) = 0$. Each sensor observation is described through the observation model $f_i^p = f_i(z_i \mid p_i)$; Figure 5.4. The features described by $\{p_i\}$ are considered to be related to some underlying global geometric description $g(x, p) = 0$ through the transformation

$$\{p_i\} = h_i(p); \qquad p_i \in \mathcal{P}_i, \quad p \in \mathcal{P}. \qquad (5.11)$$

This is a stochastic transformation of one type of geometry to another. The forward transformation $h_i(\cdot)$ is usually well-defined, and if we have some prior information about p, h_i can be used to obtain prior feature information. However, in general no prior information will be available. In this case we must use the inverse transform to generate hypotheses about the underlying geometry:

$$p = h_i^{-1}(p_i) \tag{5.12}$$

The inverse transform is usually indeterminant; a single estimate p_i, insufficient to generate a complete hypotheses p.

Consider now a multi-sensor system comprising n sensors $S_1, \cdots, S_n$, each taking a set of geometric observations $\{z_i\}$ of geometric features $\{p_i\}$ in the environment. We will assume that prior clustering of observations has been performed and that the resulting observation z_i can be considered as Gaussian $z_i \sim N(\hat{p}_i, \Lambda_i)$. From these observations, each sensor can make individual estimates $\delta_i(z_i) \in P_i$ of possible features of a particular type, resulting in a set of feature hypotheses;

$$\begin{aligned} \{p_i'\} &= \{\delta_i(z_{i,1}), \cdots, \delta_i(z_{i,1})\} \\ &= \{p_{i,1}', \cdots, p_{i,1}'\} \end{aligned}$$

Following our development of multi-Bayesian teams, we will consider the posterior distribution $f(p_i \mid z_i)$ as the preference order induced by the observations z_i on possible hypotheses p_i. With this identification, we can consider the estimates p_i' as modeled by a Gaussian $p_{i,j}' \sim N(p_{i,j}, \Lambda_{i,j})$. To transform these feature hypotheses on some underlying geometry, we need to apply the transformation $h^{-1}(\cdot)$ to the set $\{p_i'\}$. If the elements of this hypotheses set are considered Gaussian, then following Chapter 2, we can approximate this transform by a transformation of mean and variance, so that for each $\delta_i(z_i)$, we have;

$$\hat{p}' = h_i^{-1}[p_i'] = h_i^{-1}[\delta(z_i)]$$

$$\Lambda_p^{-1} = \left(\frac{\partial h_i^{-1}}{\partial p_i}\right) \Lambda_i^{-1} \left(\frac{\partial h_i^{-1}}{\partial p_i}\right)^T$$

(5.13)

Thus by transforming each feature hypothesis, we can obtain from each sensor S_i a set of hypotheses $\{p\}_i$ on the underlying environment geometry, each of which can be described by a mean and variance. The information matrix Λ_p^{-1} will be singular, no information in one or more degrees of freedom, because the transform $h_i^{-1}(\cdot)$ is indeterminant. This provides for the feature hypotheses p_i' to generate only partial estimates of the underlying geometry.

We now have a means of generating partial hypotheses of the environment geometry, regardless of the existence of prior information, in a language common to all sensors. It remains now to compare verify and combine these partial estimates to provide a complete description of the underlying environment geometry. We will formulate this problem in terms of a multi-Bayesian team, verifying individual sensor hypotheses by comparing their contribution to some global environment description.

Consider the preference placed on environment descriptions p by each sensors partial estimates p' as the posterior density;

$$f_i(p \mid p') = f_i(p \mid h_i^{-1}[\delta_i(z_i)]), \qquad i = 1, \cdots, n.$$

This preference ordering represents an *individual* sensors preferred contribution to a team consensus. We will define the *team* preference ordering as the joint posterior density on all contributions;

$$F(p \mid h_1^{-1}[\delta_1(z_1)], \cdots, h_n^{-1}[\delta_n(z_n)]) = \prod_{i=1}^{n} f_i(p \mid h_i^{-1}[\delta_i(z_i)])$$

(5.14)

We will seek consensus values of p which make $F(p \mid \cdot)$ convex.

The hypotheses generated from different sensor observations will be associated with many different values of $\mathbf{p}$. Rather than comparing all of these hypotheses in one opinion pool, it makes sense to compare different estimates pair-wise, recursively clustering hypotheses into groups, one associated with each value of $\mathbf{p}$. This pair-wise comparison can be computationally more attractive and because Equation 5.14 is order independent, construction of the joint posterior density by recursively updating the terms f_i can be justified.

From Equation 5.9, the pair-wise comparison of two hypotheses $\mathbf{h}_i^{-1}[\delta_i(\mathbf{z}_i)]$ and $\mathbf{h}_j^{-1}[\delta_j(\mathbf{z}_j)]$, requires that we can find a consensus hypotheses $\mathbf{p}$ which satisfies;

$$\left[\Lambda_i^{-1}\left(\mathbf{p} - \mathbf{h}_i^{-1}[\delta_i(\mathbf{z}_i)]\right) + \Lambda_j^{-1}\left(\mathbf{p} - \mathbf{h}_j^{-1}[\delta_j(\mathbf{z}_j)]\right)\right]$$

$$\times\left[\Lambda_i^{-1} + \Lambda_j^{-1}\right]^{-1}$$

$$\times\left[\Lambda_i^{-1}\left(\mathbf{p} - \mathbf{h}_i^{-1}[\delta_i(\mathbf{z}_i)]\right) + \Lambda_j^{-1}\left(\mathbf{p} - \mathbf{h}_j^{-1}[\delta_j(\mathbf{z}_j)]\right)\right]^T \leq 1$$

$$(5.15)$$

The value of $\mathbf{p}$ that makes the left side of this equation a minimum and is also the consensus, when it exists, is given by (Equation 5.10):

$$\overline{\mathbf{p}} = \left[\Lambda_i^{-1} + \Lambda_j^{-1}\right]^{-1}\left[\Lambda_i^{-1}\mathbf{h}_i^{-1}[\delta_i(\mathbf{z}_i)] + \Lambda_j^{-1}\mathbf{h}_j^{-1}[\delta_j(\mathbf{z}_j)]\right] \quad (5.16)$$

Substituting Equation 5.16 into Equation 5.15 gives;

$$\left(\mathbf{h}_i^{-1}[\delta_i(\mathbf{z}_i)] - \mathbf{h}_j^{-1}[\delta_j(\mathbf{z}_j)]\right)(\Lambda_i + \Lambda_j)^{-1}$$

$$(5.17)$$

$$\left(\mathbf{h}_i^{-1}[\delta_i(\mathbf{z}_i)] - \mathbf{h}_j^{-1}[\delta_j(\mathbf{z}_j)]\right)^T \leq 1$$

Equations 5.17 and 5.16 are equivalent to a pair-wise convexity analysis of Equation 5.14, they can be applied recursively to verify and combine hypotheses from different sensors in a multi-Bayesian team: Pairs of hypotheses are compared using Equation 5.17, verifying disparate sensor opinions, and recursively constructing estimates of the underlying geometry. Equation 5.16 is used to recursively combine these groups into a set of environment hypotheses.

It is important to understand the effect of interpreting the partial feature estimates provided by each sensor in terms of the hypotheses on some global environment geometry. The contribution made by a sensor to a team consensus is evaluated using a transformation of the observed feature estimates, in a language common to all team members. The transformation of observations is often only partial, so that the sensor may have invariant (or indifferent) opinion preference to certain degrees of freedom of the team hypothesis. Thus when observations from diverse sources are compared using Equation 5.17, there may be degrees of freedom over which one of the sensors will not offer an opinion, in which case the other sensor will dominate the consensus in that direction.

The recursive application of Equations 5.17 and 5.16 represent an efficient means of verifying and combing hypotheses generated by any number of diverse information sources within the structure of a Multi-Bayesian team.

5.3.3   Constraint and Coordination

Our development of the multi-Bayesian team has provided a means of generating, verifying and combing information from different sensor systems. In this section, we will consider the dynamic dependence between these sources of information and the ways in which it can be used. This dynamic element underlies the structure of a team; it is the "intelligence" or local decision making ability of sensors which provides for the fullest use of a multi-sensor system's power. There are a number of important sensing tasks in which the dynamic use of sensory information is essential:

1. When a single sensor supplies partial information we can use another sensor to "fill in the gaps" or compliment it's observations. In this case there is a dialogue or search process to find the appropriate information. A good example of this is the use of a tactile probe to fill in the parts of a scene obscured to a vision system [Allen 85].

2. Information from one sensor can often profitably be used to guide another sensor, and so reduce the number of hypotheses that must be found. For example we might want to use vergence or focus estimates of depth to restrict the search space in stereo matching [Krotkov 87].

3. Many sensors have cues that can resolve features over a number of different ranges or are in themselves multi-resolution devices. In this cases we can use the coarse resolution information sources to guide the attention of the finer resolution cues. For example, the pre-attentive process guiding the interest of fovial vision, or a pyramid of image segmentations [Mallat 87].

If we are to fully understand and utilize this dynamic ability, we must be able to describe and model the interactions that take place between different information sources. Consider two sensors taking observations $\{z_i\}$ and $\{z_j\}$ of different types of features $g_i(x, p_i) = 0$ and $g_j(x, p_j) = 0$, as in Figure 5.5. Suppose that these two sensors each observe specific features that are related to a single element of the common geometry p; $g(x, p) = 0$,. The relation between the observed features and this common element is described by the transformations $p = h_i^{-1}(p_i)$ and $p = h_i^{-1}(p_i)$. It follows that the nominal (perfect information) relation between two disparate features is just:

$$h_i^{-1}(p_i) = h_j^{-1}(p_j) \qquad (5.18)$$

Our analysis of stochastic topology in Chapter 2, further requires that any *consistent* hypotheses p_i' and p_j' of the features

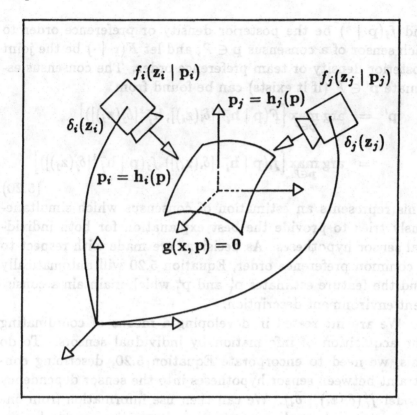

Figure 5.5: Two sensors observing different features of a common geometric object

p_i and p_j based on the observations z_i and z_j *must* also satisfy this relation;

$$h_i^{-1}[p_i' = \delta_i(z_i)] = h_j^{-1}[p_j' = \delta_j(z_j)] \qquad (5.19)$$

This means that we cannot make the decisions δ_i and δ_j independently, rather we must base our estimates on the resulting consensus element of geometry $p' \in P$, and calculate the resultant individual feature estimates from the transforms $p_i' = h_i(p')$ and $p_j' = h_j(p')$. The use of both individual and team preference orders in the multi-Bayesian system allows us to make this consistent interpretation directly: Let $f_i(p \mid \cdot)$

and $f_j(\mathbf{p} \mid \cdot)$ be the posterior density or preference order to each sensor of a consensus $\mathbf{p} \in \mathcal{P}$, and let $F(\mathbf{p} \mid \cdot)$ be the joint posterior density or team preference order. The consensus estimate $\mathbf{p}' \in \mathcal{P}$ (if it exists) can be found from

$$\mathbf{p}' = \arg\max_{\mathbf{p} \in \mathcal{P}} \left[F(\mathbf{p} \mid \mathbf{h}_i^{-1}[\delta_i(\mathbf{z}_i)], \mathbf{h}_j^{-1}[\delta_j(\mathbf{z}_j)]) \right]$$

$$= \arg\max_{\mathbf{p} \in \mathcal{P}} \left[f_i(\mathbf{p} \mid \mathbf{h}_i^{-1}[\delta_i(\mathbf{z}_i)]) \, f_j(\mathbf{p} \mid \mathbf{h}_j^{-1}[\delta_j(\mathbf{z}_j)]) \right]$$

$$(5.20)$$

This represents an estimation of consensus which simultaneously tries to provide the best explanation for both individual sensor hypotheses. As decisions are made with respect to a common preference order, Equation 5.20 will automatically find the feature estimates $\mathbf{p}_1'$ and $\mathbf{p}_2'$ which maintain a consistent environment description.

We are interested in developing a means of coordinating the acquisition of information by individual sensors. To do this, we need to encorporate Equation 5.20, describing constraint between sensor hypotheses into the sensor dependency model $f_i^\delta(\delta_i(\mathbf{z}_i) \mid \overline{\delta}_i)$. We can then use information from one source to guide and constrain the observations made by another, different source.

Consider two sensors S_i and S_j exchanging information in terms of a common geometry $\mathbf{p}$ (Figure 5.6). The information or initial hypotheses, provided by the ith sensor are described by the transformation $\mathbf{h}^{-1}[\delta_i(\mathbf{z}_i)]$, representing a set $\{\mathbf{p}\}_i$ of initial geometry hypotheses. These hypotheses are communicated to the jth sensor in terms of a set of partial means $\{\hat{\mathbf{p}}\}_i$ and singular information matrices $\{\Lambda_p^{-1}\}_i$ found from Equation 5.13. This information can then be used as prior information on the jth sensors observations:

$$f(\mathbf{z}_i, \mathbf{z}_j) = f_j^\delta(\mathbf{z}_j \mid \mathbf{h}_i^{-1}[\delta_i(\mathbf{z}_i)]) f_i(\mathbf{h}_i^{-1}[\delta_i(\mathbf{z}_i)]) \qquad (5.21)$$

This prior information can be used to restrict the search space for consensus hypotheses by guiding the observation process,

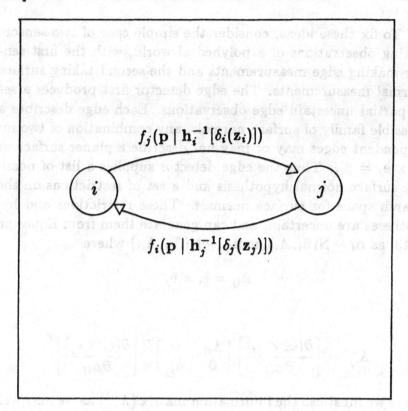

Figure 5.6: Two sensors exchanging information in terms of a common geometry

so that;

$$f(\mathbf{p}|z_j, z_i) = F(\mathbf{p} \mid \mathbf{h}_j^{-1}[\delta_j(z_j)], \mathbf{h}_i^{-1}[\delta_i(z_i)])$$

$$= f_j(\mathbf{p} \mid z_j, \mathbf{h}_i^{-1}[\delta_i(z_i)]) \, f^\delta(z_j \mid \mathbf{h}_i^{-1}[\delta_i(z_i)]) \, f_i(z_i)$$

In Equation 5.20 this reduces the consensus seeking problem to the following form:

$$\mathbf{p}' = \arg \max_{\mathbf{p} \in \mathcal{P}} f_j(\mathbf{p} \mid z_j, \mathbf{h}_i^{-1}[\delta_i(z_i)]) \qquad (5.22)$$

The advantage of coordination is now clear, the group consensus can now be found by maximization of $f_j(\mathbf{p} \mid \cdot)$ alone.

To fix these ideas, consider the simple case of two sensors taking observations of a polyhedral world, with the first sensor making edge measurements and the second taking surface normal measurements. The edge detector first produces a set of partial uncertain edge observations. Each edge describes a possible family of surface normals, any combination of two independent edges may or may not describe a planer surface as $e_i \times e_j = n_{ij}$. Thus the edge detector supplies a list of possible surface normal hypothesis and a set of restrictions on the search space for surface normals. These restrictions and hypotheses are uncertain, and can generate them from Equation 5.13, as $n_i \sim N(\hat{n}_i, \Lambda_i)$ or $n_{ij} \sim N(\hat{n}_{ij}, \Lambda_{ij})$ where

$$\hat{n}_{ij} = \hat{e}_i \times \hat{e}_j$$

and

$$\Lambda_{ij}^{-1} = \left[\frac{\partial(e_i \times e_j)}{\partial n_{ij}}\right] \left[\begin{array}{cc} \Lambda_{e_i} & 0 \\ 0 & \Lambda_{e_j} \end{array}\right]^{-1} \left[\frac{\partial(e_i \times e_j)}{\partial n_{ij}}\right]^T$$

Here we must use the information matrix (Λ^{-1}) as we may well have partial information or infinite uncertainty in some degrees of freedom. Indeed for the case $\Lambda_{e_j}^{-1} = 0$, we restrict the search space to an area which has infinite uncertainty perpendicular to the edge e_i. The information dependency model $f^\delta(\cdot)$ describes this transformation of edge observations to surface normal hypothesis and search space restrictions. The sensor that measures surface normals takes these restrictions and uses them as prior information to guide it's own observations. These observations can subsequently be used to confirm or disprove the various hypothesis $\hat{n}_{ij}$ originally supplied by the edge detector. The two sensors can now provide a robust first estimate of the polyhedra surfaces in the environment. The surface normals found by the second sensor can now be used to guide the second pass of the edge detector. This process should continue until a sufficiently accurate consensus can be reached.

5.4 Sensor Control

The ability of some sensors to move or change their state provides a powerful means to extract information about an environment. To understand how to utilize this ability, we need to know how observation hypotheses vary with changes in the sensor location or state. This analysis is particularly important in such areas as mobile robot navigation, tactile probing, or active vision.

In Chapter 3, we developed a model of state dependence in terms of a state information structure η_i^z, described by a probability distribution $f_x(\mathbf{x}_i \mid \mathbf{p}_i, \bar{\delta}_i)$. This model acts as a view-modifier, transforming prior information and expected observations into the sensors current viewing frame. We will now demonstrate how this state model can be used to develop active sensor control strategies.

Consider the example of a mobile sensor taking observations of a feature $\mathbf{p}$ in the environment. Let $\mathbf{T}_p$, $\mathbf{T}_s$ and $\mathbf{T}_z$ be the homogeneous transforms relating the feature coordinate system to base coordinates, the sensor to base coordinates and the sensor to the feature respectively, with $\mathbf{T}_p = \mathbf{T}_s \mathbf{T}_z$ as Figure 5.7. If the sensor takes uncertain observations of the feature, assumed Gaussian $\sim N(\mathbf{p}, \boldsymbol{\Lambda}_z)$, how should we locate the sensor to obtain the maximum information from each observation ? There are a number of cases when this situation might arise, when we have a mobile range finder, a camera mounted on a manipulator or a tactile sensor probing a surface.

Following our development of state models in Chapter 3, the state of the sensor $\mathbf{x}$, describing the location $\mathbf{T}_s(\mathbf{x})$, modifies the observation model as;

$$\mathbf{z}_i \sim N\left(\mathbf{h}(\mathbf{p}, \mathbf{x}), \mathbf{J}_s(\mathbf{x})\boldsymbol{\Lambda}_z\mathbf{J}_s^T(\mathbf{x})\right)$$

This describes the dependence of feature observation on sensor state as a transformation of geometry $\mathbf{h}(\mathbf{p}, \mathbf{x})$ and variance $\mathbf{J}_s(\mathbf{x})\boldsymbol{\Lambda}_z\mathbf{J}_s^T(\mathbf{x})$. Using the triangular constraint relation

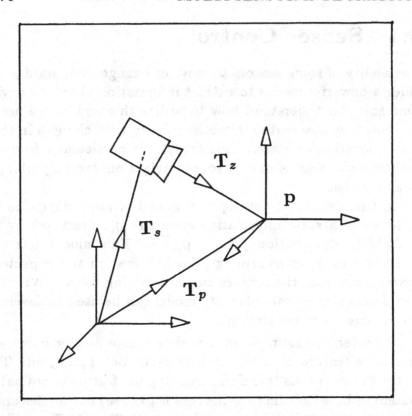

Figure 5.7: A mobile sensor taking observations of a feature in space

between base coordinates, sensor location and feature description, the modified variance in estimated object location Λ_p, can be found in terms of the observation variance Λ_z and sensor location variance Λ_s as

$$\Lambda_p = \Lambda_s(\mathbf{x}) + \mathbf{J}_s(\mathbf{x})\Lambda_z\mathbf{J}_s^T(\mathbf{x}) \qquad (5.23)$$

To gain the maximum information from an observation, we must find the sensor location $\mathbf{T}_s(\mathbf{x})$ that will minimize the elements of Λ_p (in some sense), subject to the constraints on feasible viewpoints. We make the simplifying assumption that the sensor can be located with constant variance Λ_s and that it takes observations with variance proportional to its absolute distance $|\mathbf{p}_z|$ from the object of interest: $\Lambda_z = \Lambda|\mathbf{p}_z|$. In this case any minimization of Λ_p is dependent only on $\mathbf{J}_s\Lambda_z\mathbf{J}_s^T|\mathbf{p}_z|$. If we intend to use a sensor that provides only partial information (a mobile ultrasonic ranger for example), then the minimization of variance should be considered as the maximization of the information equation:

$$\Sigma' = \mathbf{J}_s^{-T}\Sigma_z\mathbf{J}_s^{-1} \quad , \qquad \Sigma = \Lambda^{-1} \qquad (5.24)$$

Maximization of a matrix quantity is not strictly defined. Generaly, a sensor strategy will attempt to maximize some weighted sum of the elements of Σ'. If our sensor takes only partial observations, then we may wish to take a sequence of observations that successively maximize specific elements, localizing and resolving information in each degree of freedom independently. If however our sensor can obtain relatively dense information, maximizing the trace elements may be more appropriate.

Another interesting element of action control occurs between sensors that individuality take only partial observation information and must *cooperate* in the control of their locations to localize features in the environment. Hager and Durrant-Whyte [Hager 86] have recently presented an example of this type of active control. In this example, a team of sensors are required to pursue another team of evaders [Figure 5.8]. Each sensor can only take observations of depth. To find and chase an evader, the sensors must coordinate their motions, using triangulation between each other to provide a localizing strategy. The problem complexity is increased by placing objects in the environment. The team structure used by the sensors provides both global and local control, the members form an antagonistic team, able to disagree or join in the team consensus. This scenario enables us to ask a number of important questions about organizations, local decision making and global coordination. The description of coordinated sensor control is still at

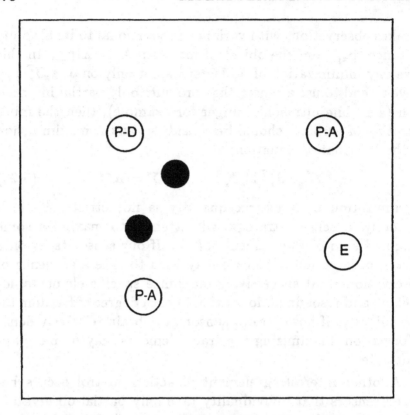

Figure 5.8: The multi-sensor pursuer-evader game

an early stage. The transportation of these ideas to multiple active sensor systems would provide an interesting basis for the analysis of problems such as vision and touch coordination.

5.5 Summary

We have developed a description of a multi-sensor system as a team of observers. The sensors are members of this team, each observing and contributing to the team-consensus view of the environment. Each individual sensor can make local decisions based on its own observations, but must cooperate with the other team members to achieve a common objective. We have described how the information structure model of sensor abilities can be used to exchange information between disparate sensor systems. This information exchange provides a basis through which individual sensors can *cooperate* with each other, resolve *conflicts* or disagreements, or *compliment* each others view of the environment. We have developed techniques to describe the coordinated acquisition of information from different sensing devices, and the control of active sensing systems. The description of a multi-sensor system as a team of decision makers provides a powerful means to understand and utilize dynamic interactions between sensing systems.

Chapter 6

IMPLEMENTATION AND RESULTS

6.1   Introduction

In this chapter, we will describe the application of the sensor modeling and integration algorithms described, to the problem of coordinating and integrating information from both visual and tactile sensing systems. This implementation was originally developed as part of a distributed robot control system [Paul 86]. The goals of this system were to provide a robust, extensible multi-sensor robot controller in which sensors and other components of the system, maintain a large degree of local autonomy or decision making capability; executing and guiding task execution in a highly parallel manner.

In this implementation, the vision system comprised a stereo camera pair mounted on a mobile head, and the tactile system, a two-fingered gripper, instrumented with tactile arrays and strain gauges, mounted on a manipulator arm. The vision system was provided with three sensory cues; a line-based stereo algorithm, a 2-D region growing algorithm, and an orientation from skewed-symmetry algorithm. The tactile-gripper, located by the manipulator, was provided with two sensory cues; an array-based region and edge algorithm, and a strain-gauge based surface normal detector. The domain of operation was restricted to a simple polyhedral world.

The purpose of this implementation was to validate the environment and sensor modeling techniques described, and to test algorithms for integrating and coordinating information from diverse sources. For this reason, the individual sensory cues were kept quite primitive, to speed processing, and the major effort was concentrated on the manipulation of the information provided by the sensing system. Each sensor algorithm was provided with an information structure describing the geometry observed, error characteristics, and dependency details. The information provided and supplied to a sensor model is described by a parameter vector and an information matrix. The transformation of these descriptions between different representations, and the integration of resulting hypotheses is handled by a coordinating mechanism. The majority of this implementation is devoted to this manipulation, coordination and integration of disparate geometric sensory information.

This implementation is not intended to be an exhaustive application of the ideas presented in previous chapters, rather indicative of procedure. In the following sections, we will describe the implementation of this hand-eye system and discuss the results obtained. Section 6.2 describes the aims and organization of the distributed robot controller. Section 6.3 discusses the scope of this system and the goals of the experimental program. Section 6.4 details the implementation of the coordination mechanism, the sensing system and the communication language. Section 6.5 describes the results of some simulation studies aimed at investigating computational problems. Section 6.6 details the experimental tests performed, describing the results obtained and problems encountered.

6.2   A Structure for Multi-Sensor Systems

We will describe a multi-sensor robot system that takes diverse geometric observations of the environment from any number of sensors and cues and integrates this information to provide a robust, consistent estimate as to the state of the robots environment. We propose that a robot system be organized as a team of expert agents implementing tactical plans and being supervised by a strategic executive. We maintain that this organization imposes sufficient structure coordination and exception handling while admitting a loosely coupled communication to be maintained during normal execution. We intend to develop a robot system that takes simple assembly tasks and implements action and sensing procedures to accomplish this goal; we are interested primarily in the coordination of intermediate level goals and not in high level planning.

We will develop an organizational structure which accounts for this uncertainty and allows the sensor agents to make mistakes or supply spurious information. The complexity of many sensor algorithms suggests that we should also distribute as much local decision making as possible to the sensor systems. This in turn allows the agents of the system to be loosely coupled, providing increased robustness and extensibility. We maintain that because of the diverse nature of sensor observations, the sensor observations the sensor agents should be allowed to maintain their own local world model, specific to local sensing requirements, and that sensors should be endowed with as much local decision making ability as possible. We propose a robot controller consisting of a distributed community of sensing, acting and reasoning expert agents communicating via a Blackboard structure [Lesser 79] to a coordinator [Figure 6.1]. The function of the coordinator is to integrate sensor information, make decisions about the allocation of tasks and

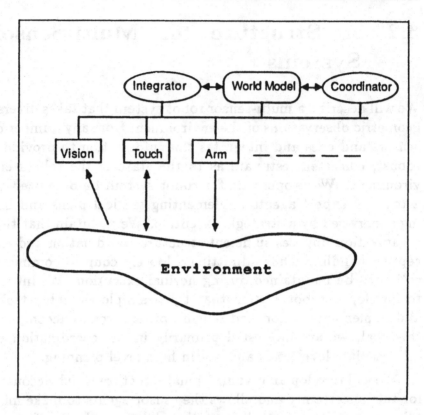

Figure 6.1: The coordinator-agent sensing system

arbitrate between the agents comprising the system. For this
the coordinator maintains a model of its best estimates of the
state of the environment and a blackboard for communicating with the other agents. Each of the distributed agents is
expert in its own localized function and contains its own localized knowledge source and agent-specific world model. We
maintain that minimal communication between team members
or agents will be achieved when distribution is performed by
function alone. That is each agent is associated with some function, vision, touch, manipulation, etc. In this way the agents
can maintain very specific functional models and algorithms,
only communicating information that is relevant to the rest of

the system.

The task domain in which we propose to work is that of mechanical part assembly. This involves the motion of one or more manipulators or objects, and the mating of such objects [Lozano-Perez 83]. Each of the actuation and sensor systems will have knowledge of the domain of operation. The coordinator will provide geometric descriptions of the following: an object, its new position, and its relationship to other objects in the work environment [Liberman 77]. Each of the subsystems will generate partial contributions to task step execution. The coordinator will synthesize a complete task step plan, going back to individual systems if necessary to request alternate plans in order to resolve conflicts between systems. Once a consistent plan is developed, the individual systems will fill in details in order to provide sensor information to the coordinator during step execution.

There are two important concepts embodied in this system; the requirement for self-knowledge and the delegation of local tasks to the sensing agents. An important characteristic of any intelligent system is a degree of knowledge about itself [Winston 81] which allows the system to reason about its capabilities. Our system comprises a network of agents each with local expertise in solving global problems. To function effectively, and avoid trivial interactions with the coordinator, the agents require a significant amount of self-knowledge to allow them to decide on what information they can supply and how best to recover from local errors. Such self-knowledge allows the coordinator to delegate much of the local work to specific agents, resulting in highly parallel problem solutions, lower processing times and an increase in overall robustness.

The coordinator's main function is to integrate the outputs of the sensing and action systems to satisfy the goal given to the system. In robot systems, like all complex structures, purposeful description is fundamental in efficient problem solving [Winston 81]. The coordinator must model the environment in a form that will allow it to integrate information from the distributed agents, form consensus views of objects and make planning decisions. The coordinator will reason in a geometric manner about objects and tasks. It is important that the coordinator have sufficient detail about a problem to enable a solution to be found, but that it is not burdened with large quantities of trivial information.

6.3   Experimental Scope

The theory developed in preceding Sections describes a methodology for the integration of partial, uncertain, disparate sensor observations. The goal of the implementation is to demonstrate the viability of these ideas within the framework of a distributed multi-sensor robot control system. We are interested to test the key elements involved in the coordination, fusion and propagation of sensory information, specifically:

1. The robustness and efficiency of the algorithms for clustering and integrating sensor observations.

2. The effectiveness of the methods for propagating sensor information and maintaining world model consistency. Of particular interest is the modeling of these mechanisms and the development of efficient representations for the object topology, allowing us to consider the issues of computational tractability, extensibility, and the possibility of highly parallel implementations.

3. The effectiveness of modeling interactions between different sensors; how this can be used to aid coordination and the active reduction of an individual sensors search space.

4. The utilization of state-dependent observation models to develop active sensing strategies, in particular; next-best

view and next-best touch control policies.

The implementation and investigation of these procedures will be developed within the framework of the coordinator-agent paradigm. The coordinator serves as the basis for the implementation of the fusion and propagation mechanisms. The world model and sensor descriptions were implemented as part of the blackboard mechanism. We maintain that the networks generated by the environment-sensor topology should be considered as an integral basis for the blackboard architecture [Durrant-Whyte 86c]. That is, the blackboard, as described by the object and sensor relations, should reflect the dynamic elements of sensor and environment interaction as an intrinsic part of its data structure. The network structure of the environment and sensors will, in the future, allow this structure to be implemented efficiently on an array processor. The implementation of the sensor agent algorithms (edge detectors for example) does not play an important part in the development of the integration and coordination algorithms. The sensor systems developed for testing the integration algorithms have been optimized for speed and local decision making capabilities. They make use of the ideas put forward for integrating single feature observations, but otherwise do not provide the same detail or accuracy of information obtained from more conventional sensor algorithms. This was not seen to be a problem, as a key idea to be tested is the ability of the integration mechanism to be robust to poor data.

The clustering, fusion and propagation algorithms were tested over a wide range of simulated and real data. The goals in these demonstrations was to test the coordinators ability to integrate observations and maintain world model consistency. The world model was maintained as an explicit structure by the blackboard, consisting of a number of object structures, the environment topology being modeled by an arbitrary number of relations between these structures. The blackboard also maintains a model of each sensors behavior characteristics. Each sensor algorithm is represented by a structure comprising the three component sensor models (observation, dependence, and state) and described by associated performance measures. These structures relate sensor input requirements to output requests. This model is coupled with a set of subroutines to translate sensor observations into the form required to provide prior information to other sensors. The performance measures describe the expected uncertainty in observations, the available density of information and the speed with which measurements can be made. The transformation subroutines implement a network of information flow between these sensors and provide an ability to reconfigure the sensors dependency model.

Models of the interaction between coarse and fine edge detectors, and interactions between edge and region-growing algorithms, were tested in limited environments. Consideration of the interactions developed between a vision system guiding a tactile probe was also investigated. The simulated data in these tests was used to provide the coordinator with a large quantity of data over long time periods. The goal of the simulations was to test the efficiency of the integration process, isolate computational bottlenecks and understand the long-term effects of information propagation on world model consistency. The sensor and cues constructed in the full implementation were limited by the sparsity of data supplied to the coordinator. The experiments conducted with these agents were designed to test the validity of the integration and propagation algorithms in real situations, over short time scales. The sparsity of data precluded the use of real data in the analysis of long term effects.

6.4 Implementation

There are three main components of the coordinator; user interface, integration mechanism, and the blackboard system (incorporating environment and sensor modeling facilities). Three agents have also been implemented: A stereo camera system, a tactile gripper mechanism and a robot location mechanism. These six elements are each configured as a process within the DPS[1] programming environment. Communication is implemented using a common protocol through the DPS port structure.

6.4.1 The Coordinator

The coordinator mechanism runs on a Vax 11/785 using a framebuffer for display purposes. At the start of operation, the world model of object descriptions and locations is loaded from a file in to the blackboard structure. The topology of relations between objects is then set up and a solid modeler displays the world model. (Figure 6.2) for example). The blackboard broadcasts a set up request to all the sensor and cue processes that have been initialized by the DPS configuration file. When acknowledgments have been received from all the sensors, the prior world model is broadcast to the agents. On receipt of the centroid locations and descriptions of the prior objects, each sensor constructs it's own local world model description. The sensors are assumed to have been previously initialized and calibrated.

Each object represented on the blackboard is described by a structure containing the object location with respect to base coordinates, a primitive shape description, a name, an information matrix and a list of objects to which it is related. The relations between objects form the world model topology. The

---

[1]DPS is a distributed programming environment developed at the University of Pennsylvannia's Grasp laboratory [Lee 85]

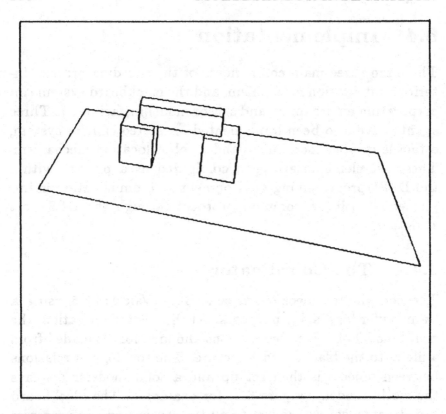

Figure 6.2: Example world model maintained by the black-board system

location of each object is described *only* with respect to the base coordinate system. The relation between any two objects can be uniquely determined by recalling that the joint variance-covariance matrix around a loop of relations is *always* singular. If an observation is made of a relation between two objects, the prior information can be generated by knowing the object locations with respect to a common coordinate system. This consideration significantly reduces the amount of information maintained by the blackboard system, and eliminates the possibility of suppling inconsistent prior information.

When the world model has been broadcast, acknowledged

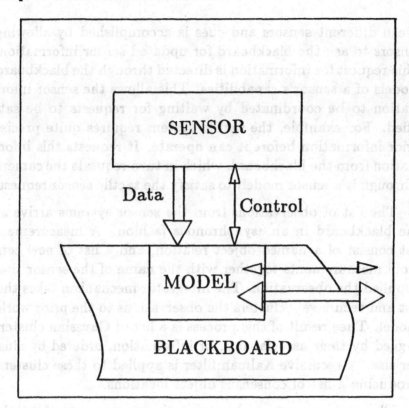

Figure 6.3: **Schematic of sensor model maintained by black-board system**

and verified by the sensor agents, the coordinator is ready to start operations. Normally requests for sensing would be selected from current task requirements. In this implementation the user places a list of objects on to the blackboard. This list of "interesting" features is broadcast to the sensor system in terms of a task, requesting information. The sensor models on the blackboard describe the sensors ability to observe different features (Figure 6.3). In the current implementation, the known environment scenario, the sensors take observations of features and update centroid locations locally, sending the new object locations to the blackboard. Communication between different sensors and cues is accomplished by allowing sensors to ask the blackboard for updated sensor information. This request for information is directed through the blackboard models of a sensor's capabilities. This allows the sensor information to be coordinated by waiting for requests to be satisfied. For example, the tactile system requires quite precise prior information before it can operate. It requests this information from the blackboard, which in turn requests the camera (through it's sensor model) to satisfy the tactile sensor request.

The list of observations from the sensor systems arrive at the blackboard in an asynchronous fashion. A measurement list consist of a named object relation, and a list of new centroid measurements together with the name of the sensor that supplied the observation. The integrator mechanism takes this list and recursively clusters the observations to the prior world model. Thee result of this process is a list of Gaussian clusters tagged by their associated prior information, ordered by cluster size. A recursive Kalman filter is applied to these clusters, producing a list of consensus object locations.

The new consensus object locations are returned to the blackboard mechanism which compares them to the current world model, integrating the result and propagating the change in consensus through the environment topology. The consensus changes in estimated object location are then broadcast to the sensor agents which subsequently update their own local environment models. This process of observation, clustering, filtering and propagation is continued until the information request has been satisfied. After each update of the blackboard world model, the coordinator's solid modeler redisplays the new environment.

This architecture is extensible to any number of new sensors, all that is required is to specify the sensor characteristics on the blackboard and to declare the machine and process locations in the DPS configuration file

Figure 6.4: The stereo camera platform

6.4.2 The Agents

The two sensor agents used in these experiments are a stereo camera platform [Fuma 86] an a tactile-force gripper mechanism. The stereo camera platform [Figure 6.4] consists of two Fairchild CCD cameras, each with computer controlled focus and aperture. These cameras are mounted on a platform with controllable pan, tilt, vergence, x and y motion. The gripper mechanism [Figure 6.5] consists of two independently driven fingers each with a Lord LTS-200 tactile sensor, comprising a 16 × 10 tactile array and six orthonormal strain gauges. The gripper has it's own controller which is supervised from an

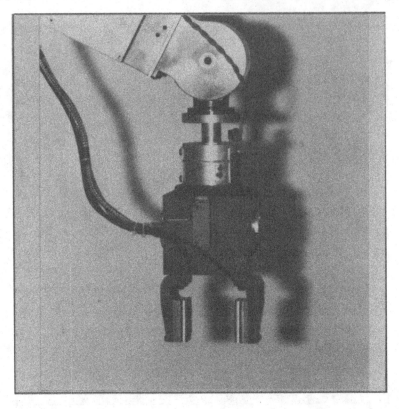

Figure 6.5: The Lord Corp. LEGS-LTS-200 tactile-gripper system

agent module located on a Vax 11/750 host computer. A Puma 500 manipulator is used to locate the gripper mechanism. End effector locations and manipulator motions are controlled by the same host as that use for the gripper.

The camera system uses three cues to extract information from the environment; a 3-D line based stereo edge detector, a 2-D region-growing procedure, and an orientation from skewed symmetry algorithm. The gripper mechanism takes measurements of locations from edges in the array image and surface normal measurements from the vector elements. The robot mechanism supplies end effector locations to transform these

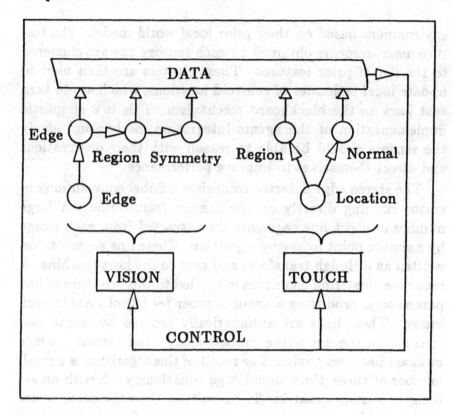

Figure 6.6: Organization of sensor systems

observations into the robot base coordinate system. The organization of this sensing system is shown in Figure 6.6.

At the start of operation, each sensor initializes it's associated hardware and loads the calibration files. When each sensor has been initialized, it sends an acknowledgment to the coordinator and receives the prior world model from the blackboard. The information is used to construct the sensors local world model. In the case of the camera and tactile sensors this is just a matter of converting the centroid and extent information in to lists of edges and surface normals. During operation, the sensor receives requests for information in terms of a list of "interesting" objects. Each sensor takes observations of the environment based on their prior local world model. The feature measurements obtained by each sensory cue are clustered to the list of prior features. These clusters are then used to update local estimates of centroid locations, which are in turn sent back to the blackboard mechanism. This is a simplistic implementation of the agents integration mechanism, ideally the sensors should be able to reason with their observations and direct themselves to improve performance.

The stereo edge detector comprises a Sobel convolution operator running directly on the camera framebuffer. A large number of short line segments are extracted from each image by a simple point following algorithm. These line segments are written as a Hough transform and sent to the host machine. A recursive clustering algorithm is applied to this sequence of line parameters, producing a small number (~ 10) of lines in each image. These lines are automatically ordered by length and strength in the clustering process and so the correspondence problem becomes trivial. The result of this algorithm is a small number of three dimensional edge equations, each with an associated variance matrix. The results of the edge detector are used to guide a region growing process of surfaces in the image [Horn 85]. Surfaces of locally uniform brightness, bounded by the edges are sought in the image plane. The surface normals are obtained by searching the gradient space within a segment, in directions restricted by the edge information. The segments and edges together with the prior information are used to estimate the angles between vertices and surfaces. This in turn allows an estimation of object orientation (in each image) due to symmetry constraints [Ballard 82]. This interaction between vision cues is described by the blackboard's model of sensor information structure.

The gripper mechanism provides edges and surfaces from the tactile array and surface normals from the vector. These are very localized, there is little doubt at this point, which object is being observed. Consequently, these observations can readily be converted in to object centroid locations, using the prior model, with respect to the end effector of the robot, transformed to base coordinates using the location recovered from the Puma controller.

6.4.3 Communication

The completed system of coordinator and agents are configured in the DPS environment. DPS makes multiple machine and process communication transparent to the user, providing a port structure through which information can be passed. This port structure is very similar to file read-write operations in UNIX-C. The implementation of communication on top of this framework was designed to allow as much modularity and flexibility as possible. A common library of functions to transfer particular data structures between named processes has been developed. Routing tables are implicitly included in this library so that the user need not be concerned about the path that a message takes to specified process. This common set of communication functions served to prevent inconsistencies between different process protocols.

All messages follow a fixed format, with a header comprising destination and source tags, message identifier, and message body. The message body resembles a Lisp-like syntax, allowing the stringing together of different commands in one message packet and enabling flexibility in future implementations. A variety of commands and message types have been programmed, including remote device control, sensing tasks and information requests.

Communication between the sensor agents and the coordinator is conducted through the blackboard. This allows for verification of sensing protocols, easy extensibility and provides a useful debugging tool. All sensor tasks follow the preset message protocol which allows for passing of location information and named interest areas.

6.5   Simulation Results

The goal of this implementation is to test the coordinators algorithms for clustering and integrating sensor observations, and to verifying the blackboard mechanism for propagating information and maintaining the consistency of the world model. There are a number of aspects of these algorithms which only become important or critical over long time periods or when many sensor observations are integrated in to the coordinators world model. These include potential bottlenecks in communication and computation, long term effects of increased world model precision, and the effect of continued distortion in the blackboard topology on consistent world modeling.

The data supplied by the currently available sensor systems is very sparse and cannot be provided in sufficient quantities to analyze these large scale effects. This is only seen as a short term limitation, and so the same coordinator mechanism was used for both simulation and experimental data. For simulation purposes, the sensor agents were replaced with simulated agents each supplying centroid observations. The simulated agents appear to the blackboard as if they are real sensor systems.

The simulated sensor agents generate observations based on their local world model. Each observation is derived from a contaminated Gaussian random vector generator. Partial information is provided by randomly zeroing rows of the observation information matrix. Spurious information is simulated by generating (with low probability) observations from an alternative world model. Each of the simulated agents randomly picks an object of interest and generates ten spurious, contaminated and partial observations, creates a list of these measurements and sends the result to the coordinator. The blackboard receives these measurement lists from all agents and directs them to the integrator. The integrator clusters and filters these observations, providing the blackboard with a consensus list of observations. The blackboard takes each element of this list and propagates this information amongst the object network. After each sequence of updates, the new world model is displayed together with a projection of the environment topology. The simulations were conducted over 30-40 sets of observations from each sensor agent.

The clustering algorithm proved to be very robust to spurious information, particularly in cases when observations were estimates of relations between objects. This is thought to be due to the constraining effect of transforming location uncertainty from a world coordinate system into an object centered coordinate frame. The recursive filtering of observations was well able to cope with partial information and was not found to cause any computational bottlenecks. However the transformation of the consensus information back to world coordinates was sensitive to orientation uncertainty. This can be appreciated by considering the effect of orientation on variance, and could probably be alleviated if the clustering and filtering operations were conducted in an object centered coordinate system. The propagation mechanism provided most of the problems in these simulations. The measurements supplied to the blackboard were transformed to world coordinates and used to calculate, recursively, updates to adjacent objects in the network. This involves transforming the information matrix from one coordinate system to another which, in the short term, causes increased sensitivity to orientation uncertainty. Thus in the initial integration cycles, the effect of orientation uncertainty significantly limits the propagation of information. However, as the orientation uncertainty is reduced the arcs of the network describing the topology provided a good mechanism for maintaining the consistency of relations between objects. During the simulations, the networks of relations between objects showed no sign of causing inconsistencies between the estimates of object locations. Again, the sensitivity to orientation could be alleviated to some extent by conducting the clustering and fusion process in an object centered coordinate system. One of the advantages of this topological network is that it provides an annealing effect, damping the distorting effects of new observations and providing a verification process by forcing consistency to be maintained.

The blackboard propagation process is the most significant computational bottleneck in the integrating process. For each observation supplied by the integration mechanism, many object locations must be updated to maintain consistency. This requires that information be transformed and propagated through the network, resulting in a large computational overhead. However, the network structure of this topology and propagation mechanism suggests that many of the consistency maintenance policies could be implemented on very parallel hardware. This could well be an important area for future consideration.

The assumed independence of observations also causes a number of important problems. It is clear that we should not be able to increase the precision of our world model arbitrarily with sensors that have only a finite resolution. In the models developed, independence of observations was assumed and consequently our algorithms are capable of providing an arbitrary high accuracy given enough sensor observations. In these simulations the only device that saved us from the consequences of this assumption was the clustering algorithm, which tended to reject more observations as the accuracy of the world model was increased. The problem with this device is that the observations now become order dependent. As previously mentioned, the only way out of this impasse is to accurately model any dependencies between sensor observations and account for correlations explicitly in the sensor models themselves.

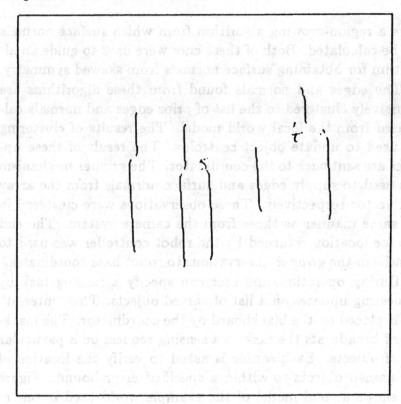

Figure 6.7: Results of edge detection on an example environment

6.6   Experimental Results

The same coordinator mechanism was used for the experimental work as was used for the simulations. The stereo camera and gripper mechanism were used as sensor agents. In the camera agent, sparse stereo edge segments were found using a framebuffer-based convolution and filtering algorithm. The results of this edge detection were converted from image to base coordinates by transforming the end points of each segments with precalculated calibration matrices [Izagurre 86]. The resulting edges are shown in Figure 6.7. These edges are used to drive a region-growing algorithm from which surface normals can be calculated. Both of these cues were used to guide an algorithm for obtaining surface normals from skewed symmetry.

The edges and normals found from these algorithms are recursively clustered to the list of prior edges and normals calculated from the local world model. The results of clustering are used to update object centroids. The result of these updates are sent back to the coordinator. The gripper mechanism was used to supply edges and surface normals from the array and vector respectively. These observations were clustered in the same manner as those from the camera system. The end effector location returned by the robot controller was used to transform the gripper observations to robot base coordinates.

During operation, the user can specify a sensing task by requesting updates on a list of named objects. This "interest" list is placed on the blackboard by the coordinator. The blackboard broadcasts the task as a sensing request on a particular set of objects. Each sensor is asked to verify the location of the named objects to within a specified error bound. Figure 6.8 shows a solid model of the example world used in one of these experiments. The camera first grabs a pair of stereo images and applies the edge detection algorithm to extract a list of stereo edge segments. These edges are then used to guide the search for surface regions and surface normals. The resulting sets of edge and normal observations were then clustered to the prior object information in each cues world model. The clustering process was used to reject any spurious observations and to supply updates to object locations. The rejection rate on stereo edges was consistently low ($\sim$ 10 per cent), this was thought to be due primarily to the small number of edges provided by the clustering algorithms, the bulk of rejections being caused by spurious stereo matching. This clearly demonstrates the utility of the gross error model for observations; providing an ability to recognize non-noise errors.

The model of dependency between the edge and surface al-

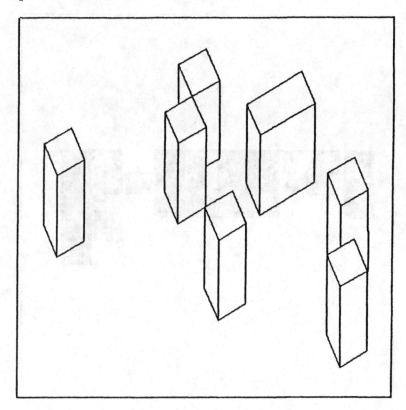

Figure 6.8: Test example of world model

gorithms converts all the edge observations into surface normal hypotheses, constraining the region growing procedure. In normal use, it is well known that surface normal algorithms usually have poor measurement characteristics, especially with uncertain lighting conditions and coarse segmentation procedures. However, using the edge detector results, interest can be localized and more detailed segmentations can be used to provide results of increased accuracy. The results from the surface normal algorithms were improved considerably by the application of the edge dependency model. Processing speeds were improved and fewer spurious normal observations were reported. Although this is only a limited dependency model, the guid-

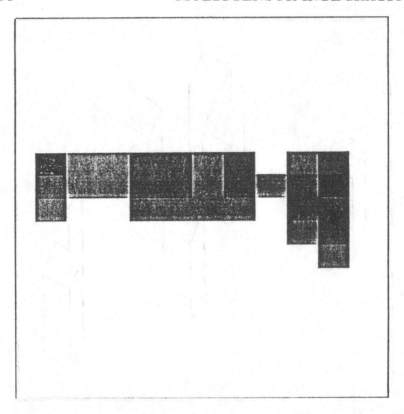

Figure 6.9: Example of a surface detected by the tactile array

ing of surface normal extraction with prior edge observations
proved to be a remarkably efficient in improving otherwise poor
measurement procedures.

The gripper takes observations of object edges and surface
normals, locations are derived from the PUMA end effector position with respect to these observations. The very local nature
of these observations precluded the possibility of mismatching
to prior information. Initially, the relative inaccuracy of the
prior world model prevents the gripper from taking observations. It takes about 2-5 camera observations before the world
prior information is sufficiently accurate to allow the gripper
to close in on an object. The locations derived from end effec-

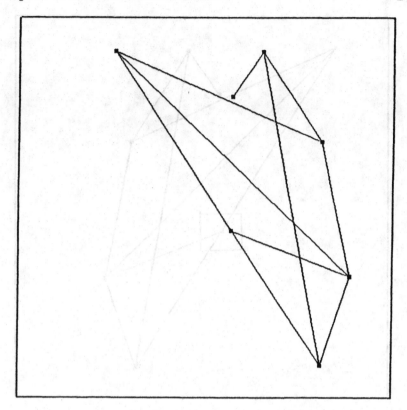

Figure 6.10: Initial world topology in x-y projection

tor locations and vector sensing were very accurate. However, the edge observations [Figure 6.9] provided by the tactile array were very poor, and in general provide almost nothing to the gripper consensus. This is primarily due to the low resolution of the tactile array, and it's relative insensitivity when compared with vector measurements. The edge and normal observations were used (in conjunction with end effector locations) to update directly the local object centroid models. No interaction model was used in this case.

On completing an observation cycle, each sensor sends a list of new object centroid observations to the blackboard. The coordinator receives the list of observations asynchronously from

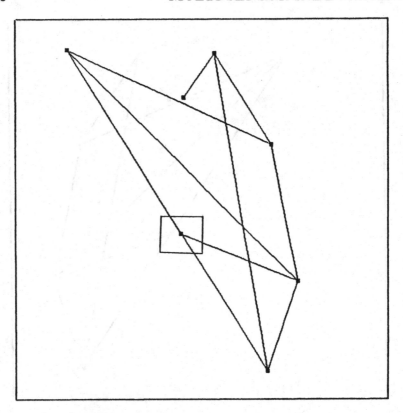

Figure 6.11: A selected area of interest

the sensor agents and supplies them, as a list, to the integrator mechanism. The integrator clusters and filters these observations to provide new consensus estimates for the world model. Very few observations were rejected in this process. This is because clustering has already been performed locally in the sensor agents. The consensus centroid locations were passed back to the blackboard and used to propagate sensor observations to provide a consistent interpretation of the environment. This mechanism provided a powerful tool with which to make maximum use of such sparse sensor observations. Figure 6.10 shows a projection in the x-y plane of the topology for the example world considered in Figure 6.8. Each node represents an

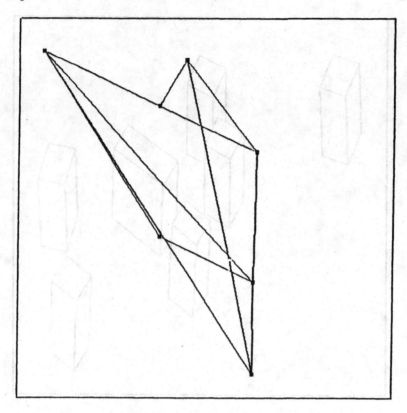

Figure 6.12: Updated world topology in x-y projection

object centroid, each arc represents a modeled relation. When an interest area is specified [Figure 6.11] and measurements are made, the integration and propagation algorithms "distort" this network in order to find a consistent interpretation for the sensor observations. Figure 6.12 shows this distortion after the integration of a single observation of one object. Note that the relations between many other objects, besides that being observed, have also changed, providing an improved estimate of their location. A solid model of the improved world model is shown in Figure 6.13. Experimental results were conducted with several other world model examples, 10-20 observations being made in each case. The results were nominally the same

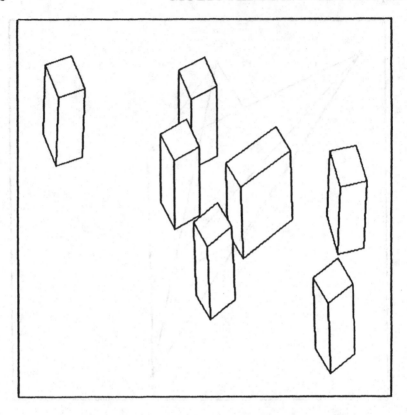

Figure 6.13: World model after update

in each case.

6.7   Summary and Conclusions

The clustering, fusion and propagation algorithms worked well for both simulated and real data sets. Of particular importance was the ability of the fusion process to use partial information and provide a mechanism for removing spurious information. The propagation algorithm proved an important structure with which to organize and maintain the world model.

The observations from the sensor systems were not as spurious as from the simulation example (except for the tactile array), however there was a tendency for sensors to drift from calibration values, causing consistent errors throughout the entire world model. In the short term this was not a problem but may become so over longer time periods. The integrator proved robust to the noisy measurements provided by the edge detector, and well able to reject spurious stereo matching errors. The simple interaction model between edge detector and region-growing algorithm proved to be a powerful method with which to localize observations. There are still however some details of the coordinator which need improving. The blackboard should also maintain a model which includes surface descriptions. This is to allow the communication of such information between different sensor systems. In addition, more consideration needs to be given to the development of effective sensor agents within this framework; more local decision making, increased resolution and range, and an ability to operate intelligent or actively to obtain information from the environment.

Chapter 7

CONCLUSIONS

7.1 Summary Discussion

We have presented the foundations for a theory of multi-sensor integration. There are two key elements to this theory; a purposeful description of a robots environment, and an effective model of a sensors ability to extract these environment descriptions.

The environment model provides a homogeneous description of uncertain geometric features and the relations or constraints between them. Methods for manipulating and transforming these uncertain geometric descriptions have been developed, providing a basis for the comparison and combination of disparate sensor observations.

The model of sensor capabilities provides a description of the observations in terms of the parameter vectors of uncertain geometric objects. This model describes the dependence of a sensors observations on prior information (observation model), other sensor's observations or actions (dependence model), and the state or location of the sensing device (state model).

Using these two models, we have developed methods for integrating multiple, uncertain, partial disparate sensor observations into a consistent, consensus description of a robot's environment. These methods are based on a probabilistic description of geometric observations, a team-theoretic description of sensors ability to extract and communicate information, and robust statistical decision making designed to cluster, integrate and propagate sensory information in a consistent manner.

We have used a team-theoretic model of dynamic sensor capabilities to investigate the coordination and control of diverse information sources. We have demonstrated that these dynamic capabilities are intrinsic to the potential power provided by multi-sensor systems. Using a multi-Bayesian formulation of the team decision problem, we have developed techniques to coordinate and control the acquisition of information by individual members of a sensory team. These techniques provide a means of describing competitive, complimentary and cooperative interactions between different information sources, and allow the development of active sensing strategies.

We have tested the viability of these techniques to the integration, coordination and control of multi-sensor robot systems, by limiting application to a system comprising of multi-cue active vision and tactile sensing.

If robot systems are ever to achieve a degree of intelligence and autonomy, they must be capable of using many different sensors in an active and dynamic manner; to resolve single sensor ambiguity, to discover and interpret their environment. We have developed a model of sensor ability in terms of a team information structure. This model has provided a powerful means of understanding the intelligent use of sensory capabilities. The further development of these models will be fundamental in endowing sensor systems with local knowledge and decision making ability.

BIBLIOGRAPHY

1. [Allen 85]
   P. Allen, "Object Recognition Using Vision and Touch",
   Ph.D Thesis, University of Pennsylvania, Philadelphia,
   1985.

2. [Allen 86]
   P. Allen, R. Bajcsy, "Two Sensors are Better Than
   One: Example of Vision and Touch", Third Int. Symp.
   Robotics Research, MIT press, 1986, p48.

3. [Ambler 75]
   A.P Ambler, R.J Popplestone, "Inferring the Positions of
   Bodies from Specified Spatial Relationships", Artificial
   Intelligence, vol 6, 1975, p157.

4. [Ambler 86]
   A.P. Ambler, "Robotics and Solid Modeling", Third Int.
   Symp. Robotics Research, MIT press, 1986, p361.

5. [Arbib 71]
   M.A Arbib, "Transformations and Somatotopy in Per-
   ceiving Systems", Int. Joint Conf. Artificial Intelligence,
   1971, p140.

6. [Ayache 87]
   N. Ayache, O. Faugeras, "Building, Registrating, and

Fusing Noisy Visual Maps", Int. Conf. Computer Vision, London, June 1987.

7. [Bacharach 75]
M. Bacharach, "Group Decisions in the face of Differences of Opinion", Management Science, vol 22, p182, 1975.

8. [Bacharach 79]
M. Bacharach, "Normal Bayesian Dialogues", J. American Statistical Soc., vol 74, p837, 1979.

9. [Bajcsy 76]
R. Bajcsy, L. Lieberman, "Texture as a Depth Cue", Computer Graphics and Image Processing, vol 5, 1976, p52.

10. [Bajcsy 84]
R. Bajcsy, "Shape from Touch", In Advances in Automation, ed. G. Saridas, JAI press, 1984.

11. [Bajcsy 86a]
R. Bajcsy, M. Mintz, E. Liebman, "A Common Framework for Edge Detection and Region Growing", U. Pennsylvania, Dept. Computer Science Tech. MS-CIS-86-13.

12. [Bajcsy 86b]
R. Bajcsy, E. Krotkov, M. Mintz, "Models of Errors and Mistakes in Machine Perception", U. Pennsylvania, Dept. Computer Science Tech. MS-CIS-86-26.

13. [Ballard 82]
D.H. Ballard, C.M. Brown, "Computer Vision", Prentice Hall, 1982.

14. [Berger 80]
J.O. Berger, "A Robust Generalized Bayes Estimator

and Confidence Region for a Multivariate Normal Mean",
The Annals of Statistics, vol 8, 1980, p716.

15. [Berger 85]
J.O. Berger, "Statistical Decisions", (second edition),
1985, Springer Verlag.

16. [Bertrand 07]
J. Bertrand, "Calcul des Probabilités", Paris 1907.

17. [Besag 74]
J. Besag, "Spatial Interaction and Statistical Analysis of
Lattice Systems", J. Royal Statistical Soc., p 193, 1974.

18. [Bickel 77]
P.J. Bickel, K.A. Doksum, "Mathematical Statistics",
1977, Holden-Day.

19. [Boissonnat 86]
J.D. Boissonnat, "An Automatic Solid Modeler for
Robotics Applications", Third Int. Symp. Robotics Re-
search", MIT press 1986, p213.

20. [Bolle 86]
R.M. Bolle, D.B. Cooper, "On Optimaly Combining
Pieces of Information, with Application to Estimating
3-D Complex-Object Position from Range Data", IEEE
Trans. Pattern Analysis and Machine Intelligence, Vol 8,
p619, 1986.

21. [Brady 82]
J.M. Brady "Computational Approaches to Image Understanding", Computer Surveys, vol 14, p3, 1982.

22. [Brady 84]
J.M. Brady, R.P. Paul, editors, "The First International
Symposium of Robotics Research", MIT press, 1984.

23. [Brady 85]
    J.M. Brady, H. Asada, "Describing Surfaces", Second Int.
    Symp. Robotics Research, MIT press 1985, p5.

24. [Brooks 81]
    R.A Brooks, "Reasoning Among 3D Models and 2D Images", Artificial Intelligence, vol 17, 1981, p285.

25. [Brooks 82]
    R.A. Brooks, "Symbolic Error Analysis and Robot Planning", Int. J. Rob. Research., vol 1,no 4, 1982, p29.

26. [Brooks 84]
    R.A. Brooks, "Aspects of Mobile Robot Visual Map Making", Second Int. Symp. Robotics Research, MIT press 1984, p287.

27. [Brooks 85]
    R.A. Brooks, "A Mobile Robot Project", MIT working Paper 265, 1985.

28. [Brooks 86a]
    R.A Brooks, "A Layered Intelligent Control System for a Mobile Robot", Third Int. Symp. Robotics Research, MIT press, 1986.

29. [Brooks 86b]
    R.A. Brooks, "A Robust Layered Control System for a Mobile Robot" IEEE J. Robotics and Automation, 1986, vol 2, p14.

30. [Cammarata 83]
    S. Cammarata, D. McArthur, "Strategies of Cooperation in Distributed Problem Solving", Int. Joint Conf. Artificial Intelligence, 1983, p767.

31. [Carbonell 81]
    J.G Carbonell, "Counterplanning: A Strategy-Based

Model of Adversary Planning", Artificial Intelligence, vol 16, 1981, p295.

32. [Chandrsekaran 81]
B.   Chandrsekaran,   "Natural and Social System Metaphors for Distributed Problem Solving", IEEE Trans. Systems Man and Cybernetics, vol 11, 1981, p1.

33. [Chartrand 77]
"Graphs as Mathematical Models", Pringle, 1977.

34. [Chatila 85]
R. Chatila, J.P Laumond, "Position Referencing and Consistent World Modeling for Mobile Robots", Proc. IEEE Int. Conf. Robotics, 1985, p138.

35. [Chatila 86]
R. Chatila, "Mobile Robot Navigation: Space Modeling and Decisional Processes", Third Int. Symposium Robotics Research, MIT press, 1986.

36. [Chu 72]
K. Chu, "Team Decision Theory and Information Structures in Optimal Control Problems Part II", IEEE Trans. Automatic Control, vol 17, 1972, p22.

37. [Corkill 79]
D.D. Corkill, "Hierachical Planning in a Distributed Environment", Int. Joint Conf. Artificial Intelligence, 1979, p168.

38. [Crowley 84]
J.L. Crowley, "Navigation for an Intelligent Mobile Robot", Carnegie Mellon University Tech. CMU-RI-TR-84-18, 1984.

39. [Crowley 86]
    J.L. Crowley, "Representation of a Composite Surface Model", Proc. IEEE Conf. Robotics and Automation, 1986, p1455.

40. [Cullingford 81]
    R.E. Cullingford, "Integrating Knowledge Sources for Computer Understanding Tasks", IEEE Trans. Systems Man and Cybernetics, vol 11, 1981, p52.

41. [Davis 83]
    R. Davis, R.G. Smith, "Negotiation as a Metaphor for Distributed Problem Solving", Artificial Intelligence, vol 20, 1983, p63.

42. [Davidson 68]
    R. Davidson, "Some Arithmetic and Geometry in Probability Theory", Ph.D Thesis, Cambridge University, 1968.

43. [Dempster 68]
    A.P. Dempster, "A Generalization of Bayesian Inference", J. Royal Statistical Soc., Series A, p 205, 1968.

44. [Dorny 75]
    C.N. Dorny, "A Vector Space Approach to Models and Optimization", John Wiley, 1975.

45. [Dresher 61]
    M. Dresher, "Games of Strategy", Prentice Hall, 1961.

46. [Duda 73]
    R. Duda, A. Hart, "Pattern Recognition and Scene Analysis", John Wiley, 1973.

47. [Dufay 84] B. Dufay, Latombe, "An Approach to Automatic Robot Programming Based on Inductive Learn-

ing", First Int. Symp. Robotics Research, MIT press 1984.

48. [Durfee 85]
E.H. Durfee, V.R. Lesser, D.D. Corkill, "Coherent Cooperation Among Communicating Problem Solvers", Proc. Workshop on Distributed Problem Solving, p231, 1985.

49. [Durrant-Whyte 85]
H.F. Durrant-Whyte, "Integration of Distributed Sensor Information: An Application to a Robot System Coordinator", Proc IEEE Int. Conf. Systems Man and Cybernetics, 1985 , p415.

50. [Durrant-Whyte 86a]
H.F. Durrant-Whyte, "Integration of Disparate Sensor Observations", Proc. IEEE Int. Conf. Robotics and Automation, 1986, p1464.

51. [Durrant-Whyte 86b]
H.F. Durrant-Whyte, "Concerning Uncertain Geometry in Robotics", Proc. Int. Workshop on Geometric Reasoning., Oxford U.K. June 1986, to appear, IEEE J. Robotics and Automation.

52. [Durrant-Whyte 86c]
H.F. Durrant-Whyte, R. Bajcsy, "Using a Blackboard Architecture to Integrate Disparate Sensor Observations", DARPA workshop on Blackboard Systems for Robot Perception and Control Pittsburg PA, June 1986.

53. [Durrant-Whyte 87a]
H.F. Durrant-Whyte, "Uncertain Geometry", IEEE Int. Conf. Robotics and Automation, p851, 1987.

54. [Durrant-Whyte 87b]
H.F. Durrant-Whyte, "Consistent Integration and Prop-

agation of Distributed Sensor Observations", Int. Journal of Robotics Research, vol 6, No 3, 1987.

55. [Durrant-Whyte 87c]
H.F. Durrant-Whyte, "Sensor Models and Multi-Sensor Integration", To appear Special Edition on Sensor Fusion, Int. J. Robotics Research, 1987.

56. [Eastman 71]
C.M Eastman, "Heuristic Algorithms for Automated Space Planning", Int. Joint Conf. Artificial Intelligence, 1971, p27.

57. [Elfes 83]
A. Elfes, S.N. Talukdar "A Distributed Control System for the CMU Rover", Int. Joint Conf. Artificial Intelligence, 1983, p830.

58. [Erdman 79]
L.D. Erdman, P.E. London, S.F. Fickas, "The Design and an Example use of HERSAY III", Int. Joint Conf. Artificial Intelligence, 1979, p168.

59. [Erdman 84]
M.A. Erdman, "On Motion Planning with Uncertainty", MIT M.Sc thesis, 1984.

60. [Erdman 85]
M. Erdman, "Using Backprojections for Fine Motion Planning with Uncertainty", Proc. IEEE Int. Conf. Robotics, 1985, p549.

61. [Fahlman 74]
S.E Fahlman, "A Planning System for Robot Construction Tasks", Artificial Intelligence, vol 5, 1974, p1.

62. [Faugeras 86]
O. Faugeras, N. Ayache, "Building Visual Maps by Combining Noisy Stereo Measurements", Proc. IEEE Conf. Robotics and Automation, 1986.

63. [Feldman 71]
J.A Feldman, R.F Sproull, "Systems Support for the Stanford Hand-Eye System", Int. Joint Conf. Artificial Intelligence, 1971, p183.

64. [Fikes 71]
R.E Fikes, N.J Nilsson, "STRIPS: A New Approach to the Application of Theorem Proving to Problem Solving", Int. Joint Conf. Artificial Intelligence, 1971, p608.

65. [Flynn 85]
A.M. Flynn, "Redundant Senors for Mobile Robot Navigation", MIT M.Sc Thesis, 1985.

66. [Flynn 87]
A.M. Flynn, "Redundant Senors for Mobile Robot Navigation", To appear Special Edition on Sensor Fusion, Int. J. Robotics Research, 1987.

67. [Fox 81]
M.S. Fox, "An Organizational View of Distributed Systems", IEEE Trans. Systems Man and Cybernetics, vol 11, 1981, p70.

68. [French 80]
S. French "Updating Belief in the Light of Someone Else's Opinion", J. Royal Statistical Society, vol 143, p43, 1980.

69. [Fuma 86]
F. Fuma, E. Krotkov, J. Summers, "The Pennsylvania Active Camera System", U. Pennsylvania Dept. Computer Science Tech. MS-CIS-86-15, 1986.

70. [Galbraith 73]
    J.R. Galbraith, "The Design of Complex Organizations",
    Addison-Wesley 1973.

71. [Garvey 81]
    T.D Garvey, J.D Lowrance, "An inference technique for
    integrating knowledge from disparate sources", Int. Joint
    Conf. Artificial Intelligence, 1981, p319.

72. [Gelb 74]
    M. Gelb, "Applied Optimal Estimation", MIT press
    1974.

73. [Geman 84]
    S. Geman, D. Geman, "Stochastic Relaxation, Gibbs Dis-
    tributions, and the Bayes Restoration of Images", IEEE
    Trans. Pattern Analysis and Machine Intelligence vol 6,
    p1984, 1984.

74. [Genest 84]
    C. Genest, "A Characterization Theorem for Externally
    Bayesian Groups" The Annals of Statistics, vol 12, p1100,
    1984.

75. [Georgeff 82]
    M. Georgeff, "Communication and Interaction in Multi-
    Agent Planning", AAAI, 1982, p125.

76. [Ghallab 84]
    M. Ghallab, "Task Execution Monitoring by Compiled
    Production Rules in an Advanced Multi-Sensor Robot",
    Second Int. Symp. Robotics, MIT press 1984, p391.

77. [Ghallab 86]
    M. Ghallab, "Coping with Complexity in Inference and
    Planning Systems", Third Int. Symp. Robotics Re-
    search, MIT press 1986.

78. [Gini 85]
   M. Gini, "The Role of Knowledge in the Architecture
   of a Robust Robot Controller", Proc. IEEE Int. Conf.
   Robotics, 1985, p561.

79. [Ginsberg 85]
   M.L. Ginsberg, "Decision Procedures", Proc. Workshop
   on Distributed Problem Solving, p42, 1985.

80. [Giralt 79]
   G. Giralt, R. Sobek, R. Chatila, "A Multi-Level Planning
   and Navigation System for a Mobile Robot", Int. Joint
   Conf. Artificial Intelligence, 1979, p335.

81. [Giralt 84]
   G. Giralt, "Research Trends in Decisional and Multi-
   Sensory Aspects of Third Generational Robots", Proc.
   Second Int. Symp. of Robotics Research, MIT press
   1984, p446.

82. [Grimson 84]
   W.E.L. Grimson, T. Lozano-Perez, "Model Based Recognition and Localization from Sparse Range or Tactile
   Data", International Journal of Robotics Research, vol
   3, p3, Fall 1984.

83. [Grimson 86]
   W.E.L. Grimson, T. Lozano-Perez, "Search and Sensing
   Strategies for Recognition and Localization of Two and
   Three Dimensional Objects", Third Int. Symp. Robotics
   Research, MIT press 1986, p81.

84. [Hager 86]
   G. Hager, H.F. Durrant Whyte, "Information and Multi-
   Sensor Coordination", Proc. Workshop on Uncertainty
   in AI, Philadelphia August 1986.

85. [Hager 87a]
G. Hager, "Information Maps for Active Sensor Control", U. Pennsylvania, Dept. Computer Science Tech. MS-CIS-87-07.

86. [Hager 87b]
G. Hager, "An Agent Specification Language", U. Pennsylvania, Dept. Computer Science Tech. MS-CIS-87-08.

87. [Hager 87c]
G. Hager, "Estimation Procedures for Robust Sensor Control", U. Pennsylvania, Dept. Computer Science Tech. MS-CIS-87-09.

88. [Hager 87d]
G. Hager, M. Mintz, "Searching for Information", Proc. AAAI Workshop on Uncertainty in AI, 1987.

89. [Hall 84]
D.J. Hall, "Robotic sensing devices", Carnegie Mellon University Tech. CMU-RI-TR-84-3, 1984.

90. [Harding 74]
E.F. Harding, D.G. Kendall, "Stochastic Geometry", John Wiley, 1974.

91. [Harmon 84]
S.Y. Harmon, D.W. Gage, W.A. Aviles, G.L. Bianchini, "Coordination of Intelligent Subsystems in Complex Robots", Proc. IEEE Conf. Applications of AI, 1984, p64.

92. [Harsanyi 77]
S. Harsanyi, "Rational Behavior and Bargaining", Cambridge University Press, 1977.

93. [Henderson 84]
T.C. Henderson, W.S. Fai, C. Hanson, "MKS: A Multi-sensor Kernal System", IEEE Systems Man and Cybernetics, vol 14, 1984, p784.

94. [Henderson 85]
T. Henderson, C. Hansen, "The Specification of Distributed Sensing and Control", J. of Robotic Systems", 1985.

95. [Henderson 87]
T. Henderson, et.al, "Workshop on Multi-Sensor Integration", U. Utah Computer Science Tech. UUCS-87-006.

96. [Hillis 85]
D. Hillis, "The Connection Machine", MIT press, 1985.

97. [Ho 72]
Y.C. Ho, K.C. Chu, "Team Decision Theory and Information Structures in Optimal Control", IEEE Trans. Automatic Control, 1972, vol 17, p15.

98. [Ho 74a]
Y.C. Ho, I. Blau, T. Basar, "A Tale of Four Information Structures", Harvard University Tech Report 657, part 1., 1974.

99. [Ho 74b]
Y.C. Ho, G. Hexner, "Redundancy in Team Problems", Harvard University Tech Report 657, part 2., 1974.

100. [Ho 80]
Y.C. Ho, "Team Decision Theory and Information Structures", Proceedings of the IEEE, vol 68, 1980, p644.

101. [Horn 86]
B.K. Horn, "Robot Vision", MIT press, 1986.

102. [Huber 81]
P.J. Huber, "Robust Statistics", John Wiley, 1981.

103. [Hummel 85]
R.A. Hummel, M.S. Landy "A Statistical Viewpoint on the Theory of Evidence", NYU tech 194, 1985.

104. [Isaacs 65]
R. Isaacs, "Differential games", John Wiley, 1965.

105. [Izaguirre 84]
A. Izaguirre, "Implementing Remote Control of a Robot Using VAL-II", U. Pennsylvania Dept. Computer Science Tech. MS-CIS-84, 1984.

106. [Izaguirre 87]
A. Izaguirre, J. Summers, P. Pu, "A New Development in Camera Calibration", To appear Int. J. Robotics Research, 1987.

107. [Kalman 69]
R.E Kalman, P.L Falb, M.A Arbib, "Topics in Mathematical Systems Theory", Mc Graw Hill, 1969.

108. [Kanade 86]
T. Kanade, C. Thorpe, "CMU Strategic Computing Vision Project", Carnegie Mellon University Tech. CMU-RI-TR-86-2.

109. [Kapur 85]
D. Kapur. J. Mundy, D. Musser, P. Narendran, "Reasoning About Three Dimensional Space", Proc. IEEE Int. Conf. Robotics and Automation, St. Louis, 1985.

110. [Kendall 63]
M.G. Kendall, P.A.P. Moran, "Geometrical Probability", Griffin, 1963.

111. [Kleinrock 85]
L. Kleinrock, "Distributed Systems", Communications of the ACM, November 1985, p1200.

112. [Koch 85]
E. Koch, "Simulation of Path Planning for a System with Vision and Map Updating", Proc. IEEE Int. Conf. Robotics, 1985, p155,

113. [Kornfeld 81]
W.A Kornfeld, C.E. Hewitt, "Scientific community metaphor", IEEE Trans. Systems Man and Cybernetics, vol 11, 1981, p24.

114. [Krotkov 84]
E. Krotkov "Construction of a Three Dimensional Surface Model", U. Pennsylvania Dept. Computer Science Tech. MS-CIS-84-40, 1984.

115. [Krotkov 86a]
E. Krotkov, "Focusing", U. Pennsylvania Dept. Computer Science Tech. MS-CIS-86-22, 1986.

116. [Krotkov 86b]
E. Krotkov, R. Kories, K. Henriksen, "Stereo Ranging with Verging Cameras", U. Pennsylvania Dept. Computer Science Tech. MS-CIS-86-86, 1986.

117. [Krotkov 87]
E. Krotkov, "Exploratory Visual Sensing for Determining Spatial Layout with an Agile Stereo Camera System", Ph.D Thesis, U. Pennsylvania, Dept. Computer Science, 1987.

118. [Kuhn 71]
H.W. Kuhn, G.P. Szego, "Differential Games and Related Topics", North-Holland, 1971.

119. [Landy 85a]
M.S. Landy, R.A. Hummel, "A Brief Survey of Knowledge Aggregation Methods", NYU tech 177, 1985.

120. [Landy 85b]
M.S. Landy, R.A. Hummel, "A Statistical Viewpoint on the Theory of Evidence", Courant Institute Tech 194, 1985.

121. [Lee 85] I. Lee, S. Goldwasser, "A Distributed Testbed for Active Sensory Processing", Proc. IEEE Conf. Robotics and Automation, 1985, p925.

122. [Lesser 79a]
V.R. Lesser, D.D. Corkill, "The Application of Artificial Intelligence Techniques to Cooperative Distributed Processing", Int. Joint Conf. Artificial Intelligence, 1979, p537.

123. [Lesser 79b]
V.R. Lesser, L.D. Erman, "A Retrospective View of the HERSAY II Architecture", Int. Joint Conf. Artificial Intelligence, 1979, p790.

124. [Lesser 81]
V.R. Lesser, D.D Corkill, "Functionally Accurate Cooperative Distributed Systems", IEEE Trans. Systems Man and Cybernetics, vol 11, 1981, p81.

125. [Liberman 77]
L. Liberman, M. Wesley, "AUTOPASS: an Automatic Programming System for Computer Controlled Mechanical Assembly", IBM J. Research Development, 21-4, 1977, p321.

126. [Lidley 79]
D.V. Lindley, A. Tversky, R.V. Brown, "On the Recon-

ciliation of Probability Assessments", J. Royal Statistical Society, vol 142, p146, 1979.

127. [Lowe 85] D.G. Lowe, "Perceptual Organization and Visual Recognition", Kluwer Academic Press, 1985

128. [Lozano-Perez 84]
T. Lozano-Perez, M. Mason, R. Taylor, "Automatic Synthesis of Fine Motion Strategies for Robots", Proc. First Int. Symp. of Robotics Research, MIT press, 1984, p65.

129. [Mallat 87]
S. Mallat, "A Theory for Multi-Resolution Signal Decomposition", U. Pennsylvania, Dept Computer Science Tech. MC-CIS-87-22, 1987.

130. [Marroquin 85a]
J.L. Marroquin, "Optimal Bayesian Estimators for Image Segmentation and Surface Reconstruction", MIT AI Memo 839, 1985.

131. [Marroquin 85b]
J.L. Marroquin "Probabilistic Solution of Inverse Problems", MIT AI Tech Report 860 and MIT Ph.D Thesis, 1985.

132. [Marshack 72]
J. Marshak, R. Radnor, "The Economic Theory of Teams", Yale University Press, 1972.

133. [McArthur 82]
D. McArthur, R. Steeb, "A Framework for Distributed Problem Solving", AAAI, 1982, p181.

134. [McGuire 72]
C.B. McGuire, R. Radnor, Eds., "Decision and Organization", North Holland, 1972.

135. [McKendall 87]
R. McKendall, M. Mintz, "Models of Sensor Noise and Optimal Algorithms for Estimation and Quantization in Vision Systems", U. Pennsylvania Dept. Computer Science Tech. MC-CIS-87, 1987.

136. [Minsky 72]
M. Minsky, S. Papert, "AI progress report", MIT AI Memo 252, 1972.

137. [Mintz 85] M. Mintz, "The Contaminated Gaussian", Personal Communication 1985.

138. [Mitche 86]
A. Mitche, J.K. Aggarwal, "Multiple Sensor Integration Through Image Processing: A Review", Optical Engineering, vol 23, p 380, 1986.

139. [Morris 77]
P.A. Morris, "Combining Expert Judgements: A Bayesian Approach", Management Science, vol 23, p679, 1977.

140. [Morris 84]
P.A. Morris, "An Axiomatic Approach to Expert Resolution", Management Science, vol 29, p24, 1984.

141. [Muson 71]
J.H. Munson, "Robot Planning, Execution, and Monitoring in an Uncertain Environment", Int. Joint Conf. Artificial Intelligence, 1971, p338.

142. [Nash 50]
J.F. Nash, "The Bargaining Problem", Econometrica, p155, 1950.

143. [Nilsson 80]
N.J Nilsson, "Principles of Artificial Intelligence", Tioga press, 1980.

144. [Orlando 84]
N.E. Orlando, "An Intelligent Robotics Control Scheme", American Control Conference, 1984, p204.

145. [Papoulis 65]
A. Papoulis, "Probability, Random Variables and Stochastic Processes", McGraw Hill, 1965.

146. [Paul 81]
R. Paul, "Robot Manipulators", MIT press, 1981.

147. [Paul 86]
R.P. Paul, H.F. Durrant-Whyte, Max Mintz, "A Robust Distributed Robot Control System", Third Int. Symposium of Robotics Research, MIT press 1986.

148. [Pavlin 83]
J. Pavlin, "Predicting the Performance of Distributed Knowledge-Based Systems", AAAI, 1983, p314.

149. [Pearl 82]
J. Pearl, "Reverend Bayes on Inference Engines: A Distributed Hierachical Approach", AAAI, 1982, p133.

150. [Pentland 85]
A.P. Pentland, "Perceptual Organization and the Representation of Natural Form", SRI Tech Note 357, 1985.

151. [Peterson 77]
J.L. Peterson, "Petri Nets", Computing Surveys, vol 9, p223, 1977.

152. [Poggio 84]
T. Poggio, V. Torre, "Ill-Posed Problems and Regularization Analysis in Early Vision", MIT AI Memo 773, 1984.

153. [Poggio 85]
T. Poggio, C. Koch, "Ill-Posed Problems in Early Vision: from Computational Theory to Analogue Networks", Proc. Royal Society, B 226, p303, 1985.

154. [Poincaré 12]
"Calcul des Probabilités", Paris, 1912.

155. [Popplestone 80]
R.J Popplestone, A.P Ambler, "An Interpreter for a Language Describing Assemblies", Artificial Intelligence, vol 14, 1980, p79.

156. [Pugh 85]
A. Pugh (ed.), "Robot Sensors Volume I: Vision", Springer Verlag, 1985.

157. [Pugh 86]
A. Pugh (ed.), "Robot Sensors Volume II: Tactile and non-Vision", Springer Verlag, 1986.

158. [Requicha 80]
A.A.G. Requicha, "Representations for Rigid Solids: Theory, Methods and Systems", Computing Surveys, vol 12, 1980.

159. [Rosenschein 82]
J.S. Rosenschein, "Synchronization of Multi Agent Plans", AAAI, 1982, p115.

160. [Ruoff 85]
C.F. Ruoff, "Device Organization in Advanced Robot Systems", Recent advances in robotics, 1985.

161. [Santalo 76]
L.A. Santalo, "Integral Geometry and Geometric Probability", Addison Wesley, 1976.

162. [Savage 54]
L.J. Savage, "The Foundations of Statistics", John Wiley, 1954.

163. [Segen 82]
J. Segen, A.C. Sanderson, "Model Inference and Pattern Discovering by Minimal Representation Method", CMU-RI-82-2, 1982.

164. [Shafer 85]
S.A. Shafer, "Shadows and Silhouettes in Computer Vision", Kluwer Academic Publishers, 1985.

165. [Simon 62]
H.A. Simon, "The Architecture of Complexity", Proc. Amer. Phil. Soc., vol 106, p 467, 1962.

166. [Smith 79]
R.G. Smith, "A Framework for Distributed Problem Solving", Int. Joint Conf. Artificial Intelligence, 1979, p836.

167. [Smith 80]
R.G. Smith, "The Contract Net Protocol: High Level Communication and Control in a Distributed Problem Solver", IEEE Trans Computers, vol 29, 1980, p1104.

168. [Smith 81]
R.G. Smith, R. Davis, "Frameworks for Cooperation in Distributed Problem Solving", IEEE Trans. Systems Man and Cybernetics, vol 11, 1981, p61.

169. [Smith 85]
R.G. Smith, "Report on the 1984 distributed AI workshop", AI Magazine, Fall 1985.

170. [Smith 86]
R.C. Smith, P. Cheesman, "On The Representation and Estimation of Spatial Uncertainty", Int. J. Robotics Research, Vol 5, p56, 1986.

171. [Smitley 84]
D. Smitley, "Pixel Based Stereo Vision", M.Sc Thesis, U. Pennsylvania, Dept. Computer Science, 1984.

172. [Spiegelhalter 86]
D.J. Spiegelhalter, "Probabilistic Reasoning in Predictive Expert Systems", Uncertainty in AI, North Holland Press, 1986.

173. [Stansfield 85]
S.A Stansfield, "Primitives, Features and Exploratory Procedures: Building a Robot Tactile Perception System", Proc. IEEE Conf. Robotics and Automation, 1985, p1274.

174. [Stansfield 87]
S.A. Stansfield, "Visually-Aided Tactile Exploration", U. Pennsylvania Dept. Computer Science Tech. MS-CIS-87-06.

175. [Stepankova 76]
O. Stepankova, I.M Havel, "A Logical Theory of Robot Problem Solving", Artificial Intelligence vol 7, 1976, p129.

176. [Taylor 76]
R. Taylor, "A Synthesis of Manipulator Control Programs from Task level Specifications", Stanford University Ph.D Dissertation AIM-282, 1976.

177. [Trezopoulos 82]
D. Terzopoulos, "Multi-Level Reconstruction of Visual Surfaces", MIT AI Memo 671, 1982.

178. [Trezopoulos 85]
D. Terzopoulos, "Multi-Resolution Reconstruction of Visual Surfaces", MIT Ph.D Thesis.

179. [Trezopoulos 86]
D. Terzopoulos, "Integrating Visual Information from Multiple Sources", In From Pixels to Predicates A.P. Pentland Ablex Press, 1986.

180. [Traub 85]
J.F. Traub, "An Introduction to Information-Based Complexity", Columbia University Tech, 1985.

181. [Von Neumann 43]
J. Von Neumann, O. Morgenstern, "The theory of Games and Economic Behavior", Princton Press, 1943.

182. [Weerahandi 81]
S. Weerahandi, J.V. Zidek, "Multi-Bayesian Statistical Decision Theory", J. Royal Statistical Society, vol 144, p85, 1981.

183. [Weerahandi 83]
S. Weerahandi, J.V. Zidek, "Elements of Multi-Bayesian Decision Theory", The Annals of Statistics, vol 11, p1032, 1983.

184. [Wesson 81]
R. Wesson, F. Hayes-Roth, "Network Structures for Distributed Situation Assessment", IEEE Trans. Systems Man and Cybernetics, vol 11, 1981, p5.

185. [White 84]
C.C. White, A.P. Sage, "A Model of Multiattribute Decision Making and Trade-off Weight Determination under Uncertainty", IEEE Trans Systems man cybernetics, vol 14, 1984, p223.

186. [Whitney 82]
D.E Whitney, E.F Junkel, "Applying Stochastic Control Theory to Robot Sensing Teaching and Long Term Control", Proc Amer. Control Conf., 1982.

187. [Winkler 81]
R.L. Winkler, "Combining Probability Distributions From Dependent Information Sources", Management Science, vol 27, p479, 1981.

188. [Winston 81]
P.H. Winston, "The MIT Robot", In Machine Intelligence , p431, 1981.

189. [Winston 83]
P.H. Winston, "Artificial Intelligence", Prentice Hall, 1983.

190. [Woziskowski 85]
H. Wozniskowski, "A Survey of Information-Based Complexity", Columbia University Tech, 1985.

191. [Zeytinoglu 85]
M. Zeytinoglu, M. Mintz "Optimal Fixed Size Confidence Procedures for a Restricted Parameter Space", Annals of Statistics, vol 12, 1985, p945.